日本设计 Nihon Sekkei

中国建筑工业出版社
CHINA ARCHITECTURE & BUILDING PRESS

创造可持续发展的环境与新价值观念

在社会、经济与环境激烈变革的今天，日本设计一直坚守自己的企业理念和基本价值观。我们的目标不仅仅是创造建筑环境，更重要的是为客户创造更高的价值。

日本设计是一家强调作品的个性化、多样化的设计集团。针对现代建筑所面临的世界性问题我们随时可以作出全面、迅速的反应，并运用丰富的知识、技术经验以及敏锐的感官去迎接各种挑战。

早在1968年我们打造了日本第一座超高层建筑，从那时起我们从未停止过在建筑领域的创新和开拓。

2008年，为了更好地引领绿色环保和可持续发展环境的建设，我们成立了 CEDeMa（环境创造管理中心）。

今后，作为值得信赖的合作伙伴，我们将一如既往为客户创造更高的价值而努力。

Building Value and Sustainability for the future

In a time of tumultuous social, economical and environmental changes, we need to keep focus on our aspirations and not to lose our sense of Value. At Nihon Sekkei, we understand that our business is not only the creation of the Built Environment, but more importantly the creation of Value for our clients.

We have a commitment to our clients to always come out as winners, whatever the economic climate. In the last 40 years of our history, we have learnt to adapt to and confront economic crisis and recessions in a fragile Japanese economy. Thanks to our experience, we have also diversified our business and gained the knowledge, skills and expertise to solve any design problem, in any market sector, in any part of the world. Nihon Sekkei designed the first skyscraper in Japan in 1968 and since then we have not halted to innovate and break new ground in the field of architecture. In 2008, we established CEDeMa (Center of Environmental Design and Management) as part of our initiative to lead the industry towards the creation of greener and more sustainable environments.

With Value and Sustainability as our goals, we can assure you that as your partner, together we will be able to build a better building and a better world.

目录 Content

CBD综合体城市设计	CBD Complexes Facilities	004
文化媒体建筑	Culture and Broadcasting Media Facilities	024
学校教育建筑	Educational Facilities	060
酒店办公建筑	Hotels and Offices Buildings	078
商业建筑	Commercial Facilities	110
政府行政建筑	Government Buildings	140
医疗建筑	Medical Facilities	152
住宅项目	Residential Facilities	172
城市交通综合体	Urban Transit Complex	184
城市核心空间规划	Urban Planning Projects	204

006 天津滨海新区MSD泰达广场复合开发项目
TEDA MSD Complex Development Project, Tianjin

012 上海漕河泾综合体
Cao He Jing Complex Development Project, Shanghai

018 日本新宿商务综合区及主要建筑
Shinjuku CBD and Main Buildings, Japan

CBD Complexes Facilities
CBD综合体城市设计

天津滨海新区MSD泰达广场复合开发项目
TEDA MSD Complex Development Project, Tianjin

总建筑面积	505,000 ㎡
设计期间	2008/6~2010/5
建设期间	2009/1~
竣工年月	2011/12
配合设计	天津市设计院
获奖经历	国际设计竞赛一等奖

位于天津市新CBD中心的建筑面积达50万㎡的大型综合开发项目。项目由4座130m高的金融办公楼、配套商业设施和中心公园组成。4栋塔楼分别位于基地的东西两侧，以宏伟的体量勾勒出了CBD的天际轮廓线，从视觉上凸显CBD的中心地位。由4栋塔楼围合而成的地块中心为巨大的开放空间，连接着南侧的中心公园和北侧的文化建筑群和绿化带。塔楼的高度和立面与开发区内周边的高层建筑群保持和谐统一，中央的开放空间与周边的广场、绿化带连为一体，形成了超越基地范围的城市骨架。

4栋办公楼采用CFT柱、钢梁，实现了无柱的超大办公空间，通过灵活的分割和组合，可以满足大型企业的办公要求。裙房积极采用垂直绿化，并结合下沉广场、起伏的地形，强调公园与建筑的整体感。造型独特而线形优美的中庭散落于立体公园，而每一个中庭又是以花、水、光等为主题的庭院。这些中庭在室外是休息空间，对于内部空间又可以作为导入自然光线和通风的设施。裙房内部墙面呈优美的曲线，中庭的设置消除了室外与室内的隔阂，使室内环境融入到建筑外部的公园环境中。中心公园的地形如缓缓的丘陵与裙房相连，如飘带一样的中央步行街纵贯南北。

在设计中积极采用了节能环境设计。

超高层办公楼：
- 根据立面工程学实现降低热负荷、节能的目的。北侧采用双层幕墙达到隔热、保温的目的，东西两侧设置可以遮挡太阳光的纵向百叶。
- 利用建筑屋顶的空间设置太阳能发电板。

裙房商业设施：
- 利用大面积的屋顶绿化，减少热辐射带来的影响。
- 利用垂直绿化百叶，减低热辐射。
- 利用屋面的挑天窗设置太阳能发电板。
- 利用天窗的自然采光。

Tianjin TEDA MSD complex development project is a large-scale comprehensive development project, located in the New CBD center in Tianjin with construction area of 500,000 square meters. This project consists of four 130m financial office buildings, supporting commercial facilities and Center Park. The four towers are located on the east and west side of the base; the skylines of the CBD is outlined by magnificent mass; the center of CBD is highlighted visually. The block center enclosed by the four towers is a huge open space, connecting the Central Park in south and the cultural buildings and greenery in the north. The height and elevation of the towers shall maintain harmonious and united with the surrounding high-rise buildings in the development area; the central open space and the surrounding square and vegetation belt are united as a single entity, forming an urban framework beyond the range of base.

By adopting CFT columns and steel beams, large column-free office space can be realized. Through flexible segmentation and combination, we can meet the office requirements of large enterprises.

Vertical vegetation is adopted actively in podiums; emphasize the overall sense of parks and buildings with the sunken plaza and undulating terrain. The unique style and beautiful patio is scattered in the three-dimensional park; each patio is a garden themed as flowers, water, light and others. These patios are resting rooms in the outdoor space and facilities for importing natural light and ventilation in the interior space. The internal walls of podium are beautiful curves. The patio settings eliminate the gap between outdoor and indoor, so that the indoor environment is integrated into the park environment of outside building. The terrain of Central Park is just like hilly slowly connected with the podium, the central pedestrian street like streamers stretches from north to south.

Energy-saving and environmental design is actively used in the design:
Super high-rise office building:
- According to the facade engineering, we can achieve the purpose of lowering heat load and energy-savings. Use double walls in the north side to achieve the purposes of insulation and thermal retardation; the vertical blinds are set in the east and west to block sunlight.
- Use the space of construction roof for solar panels.

The commercial facilities of Podium:
- Use a large area of vegetation roof to reduce the impact of thermal radiation.
- Use vertical vegetation blinds to reduce heat radiation
- Use the roof skylight to set solar panels.
- Use the natural light from skylight.

Central Park:
- Use the underground parking lot of "solar chimney" as a natural ventilation system.

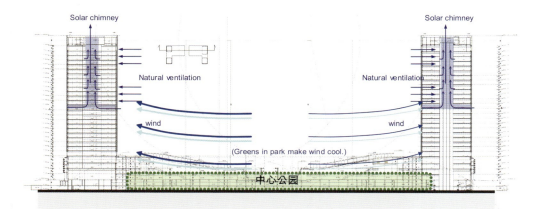

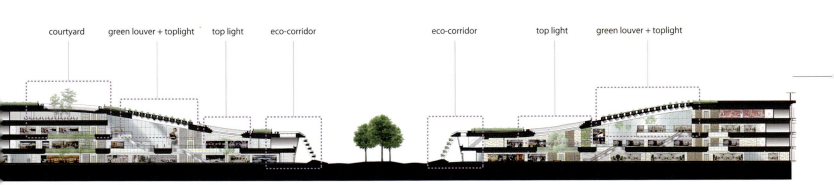

上海漕河泾综合体
Cao He Jing Complex Development Project, Shanghai

总建筑面积	154,997㎡
设计期间	2006/5~2008/5
建设期间	2008/7~2009/12
竣工年月	2009/12
配合设计	上海建筑设计研究院
摄影	林铭述摄影工作室

| 上 | 海漕河泾新兴技术开发区作为未来上海都市圈高科技园区城中城，有必要充分体现现代化、高科技特色，力求突破现有一般城市的固有模式，创造适合于漕河泾新兴技术开发区特定环境条件下的新模式。体现整体性和共享性，体现超前意识和国际性，体现城市发展的和谐性。充分融入人与自然和谐共生的理念，以现代、时尚、生态、高科技为概念特征，积极营造出一个现代化、国际化、智能化、集群化的高档综合商务区。

1. 设计理念追求高层建筑的地标性和群体建筑的统一性
- 作为高层建筑，通过简洁造型保证合理功能，鲜明的流线烘托出现代的氛围。
- 个性鲜明而具有良好统一感的造型，与聚集了高档商务楼的总部基地相匹配。
- 作为总部基地，足够的建筑间距保证了良好的室外环境以及从窗口眺望的视觉景观。

2. 营造生态绿化环境
- 健身设施的屋顶绿化显示其与周围环境和谐共存的设计思想。
- 在总部基地区域，绿化与水体景观形成了娴静商务环境。
- 会议中心与绿化融为一体，形成生态型的会议健身环境。

3. 构筑地下空间网络
- 下沉广场的舞台与瀑布激发了人们的活力。
- 总部基地的下沉广场为午休的人们提供安逸的休息场所。

4. 步行空间连接下沉广场，空间变化丰富。
庭园作为具有自然属性的空间，由绿色的花坛和各种花卉草木、水面组成，是人们散步、休闲的最佳场所，能使人们放松、充分享受自然空间。树木采用多个树种，随着季节的变化呈现出不同的表情，营造出更接近自然环境的外部空间。

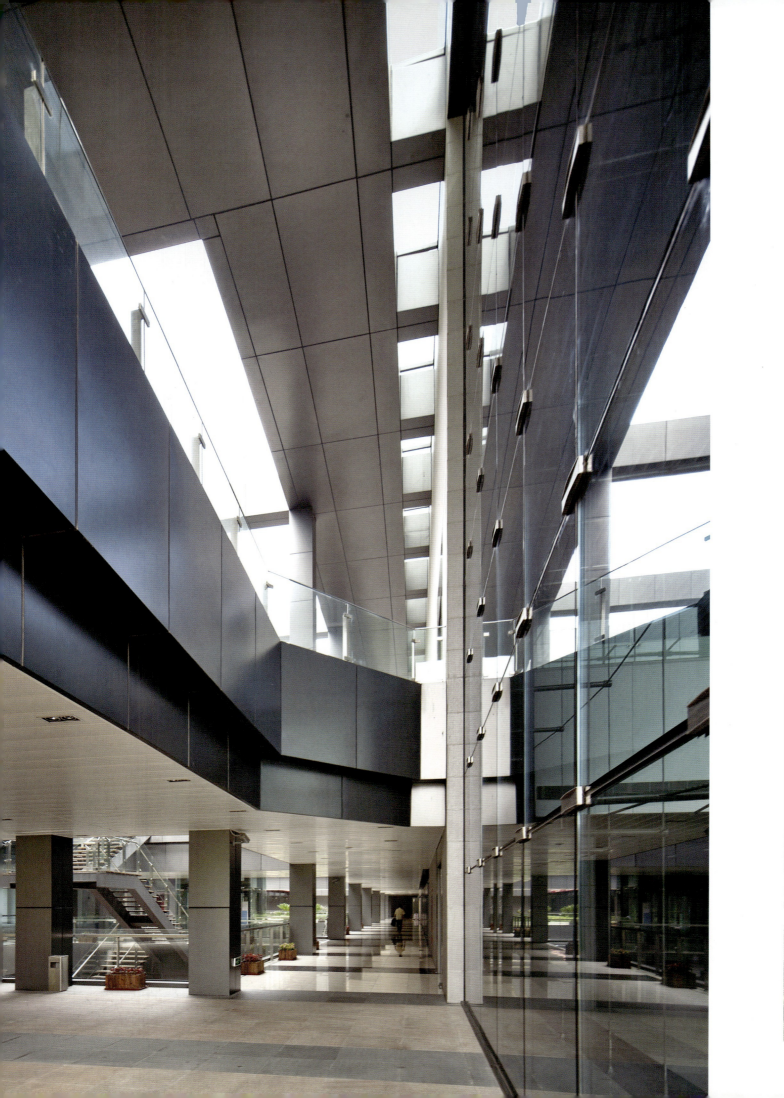

As a city within city in the future Shanghai high-tech Park, it is necessary for the Architectures in its gathering area of modern service industry to fully reflect modernity and high-tech features, strive to break through the existing inherent pattern of general city, to create new models which is suitable for the specific environmental conditions in Caohejing New Technology Development Area. It reflects integrity and sharing, forward-thinking and internationality, the harmony of urban development. The concept of harmonious coexistence between man and nature is fully integrated, featured by modernity, fashion, ecology and high-tech; we will actively create a modern, international, intelligent, high-end and clustering integrated business area.

1. The design concept pursues landmark of high-rise buildings and the unity of group buildings.
- As a high-rise building, to ensure a reasonable functional form through simple, clear flow lines express a modern atmosphere.
- The style with distinctions and good uniform matches with the headquarters base where high-end commercial buildings are gathered.
- As the headquarters base, sufficient distance between buildings ensures a good outdoor environment and visual landscape from windows.

2. To create an ecological vegetation environment
- The vegetation roof of fitness facilities demonstrates the design thought of harmonious coexistence with its surrounding environment.
- In the base region of headquarters, vegetation and water landscape form a demure business environment.
- Conference Center and Vegetation are merged, and an ecological conference and fitness environment is formed.

3. Build the network of underground space
- The sunken plaza stage and waterfall stimulate people's vitality.
- The sunken plaza in headquarters base provides a comfortable place for people who want a afternoon nap to rest.

4. The sunken plaza connects with the pedestrian space with varied spaces.
   The yard, as a space of natural property, is composed of green flower terraces and a variety of flowers, grass, woods and water. It is the best place for people to stroll and leisure; it can make people feel relaxed and fully enjoy the natural space. With multiple species, trees show different facial expressions with season changes to create an outer space which is more close to the natural environment.

日本新宿商务综合区及主要建筑
Shinjuku CBD and Main Buildings, Japan

在西新宿副都心区，日本设计参与了9栋建筑的设计，所涉及的用地范围共计有10hm²。建筑设计需要阅读和领悟规划用地的特征，并以形态表现出与周边环境和谐的关系，这意味着"日本设计"通过这9栋建筑的设计，塑造出了西新宿的城市景观。从1971年竣工的京王广场饭店，到西新宿6丁目第6地区的规划设计，"日本设计"跨越40年的奋斗成果，折射出日本城市建设的发展模式。

1960年3月出台了在淀桥净水厂的原址上"建设新宿副都心的基本方针"，同年6月确定新宿副都心建设规划之后，财团法人新宿副都心建设公社和东京都开始了基础配套设施的建设。1968年11月土地分售完毕之后，由12家土地所有人组成的新宿新都心开发协议会（SKK），根据特定街区制度提出了在绿色环绕的公开空地上，建设超高层建

摄影：三轮晃久摄影研究所

筑鳞次栉比的城市景观构想，并以"生机盎然的人性化空间的创造"为主题，制定了行人动线与车辆动线的完全分离、区域供冷供暖系统的采用、公共停车场的设置等3个主要原则，同时以此为基本方针，制定了250m的建筑限高、立面覆盖率等设计要求，以确保天空可视率。这一时期为日本城市建设的重要讨论时期，在日本城市发展史中涂抹上浓墨重彩的一笔，但是这一时期城市建设的思想及技术背景却鲜为人知。

此后在西新宿超过10栋的超高层大楼陆续建成，到1991年东京都厅搬迁时，新宿新都心开发协议会（SKK）的副都心区建设才宣告结束。这些超高层建筑形成一个建筑群，并作为象征西新宿副都心的城市地标而广为人知。

其后，每天有350万人聚散的新宿站的巨大能量推动城市的开发进程逐步扩大到副都心的周边街区。从1975年开始，日本设计开始参与由200多名土地所有人发起的新宿爱之岛的再开发建设，并几经波折，于1995年竣工。

2002年竣工的新宿橡树城（oak city）以及目前正在建设中的西新宿6丁目第6区等，这一时期的建筑物以办公、商业、住宅的综合开发为中心。这些超高层建筑虽然属于SKK区的系列建筑群，但是每栋建筑的设计都经过了与土地所有人的反复沟通及长时间的创作过程，拥有各不相同的成长史。城市建设的组织形式从初期特定街区内的建设形式不断发展，反映着特定的时代需求。

京王饭店（1971）Keio Plaza Hotel

摄影：川澄建筑摄影事务所

工学院大楼（1995）Koga Kuin University

摄影：Technical Art Ltd

新宿中央公园大厦（2011）Shinjuku Central Park City

摄影：川澄建筑摄影事务所

新宿三井ビル (1974) Shinjuku Mitsui Tower
撮影：川澄建築写真事務所

新宿哈蒙尼大厦（1997）Shinjuku Harmony Square
摄影：川澄建筑摄影事务所

新宿莫里斯大楼（1995）Shinjuku Monolith Tower
摄影：川澄建筑摄影事务所

Nihon Sekkei has been involved in the design of buildings in 9 locations on a total site area of 10ha in the emerging business area of Nishi Shinjuku. In architectural design it is always necessary to determine the characteristics of the location and realize a good relationship with the surrounding environment. In this sense we have formed the cityscape of Nishi Shinjuku through our buildings at 9 locations. Looking back over the achievements of Nihon Sekkei over the past 40 years, starting from the completion of the Keio Plaza Hotel Tokyo in 1971 up to the planning of Nishi Shinjuku 6-Chome Nishiroku Zone, it is possible to see the history of urban development in Japan.

In March 1960 the "Basic Policy for Shinjuku New Town Construction" for the site of the old Yodobashi Purification Plant was finalized, and after its establishment in June of that year, construction of the infrastructure facilities was commenced by the Shinjuku New Town Development Corporation and the Tokyo Metropolitan Government. After division of the land into lots, the 12 companies that owned or leased the land formed the Shinjuku Shin Toshin Kaihatsu Kyogikai (SKK: Shinjuku New Town Development Association). Using the special districts framework for development, they outlined a cityscape concept for a group of high-rise buildings with rich green open spaces in the center.

Based on the theme "Creation of lively human spaces", a policy was established based on three pillars: (1) complete separation of people and vehicles, (2) adoption of district heating and cooling, and (3) provision of public car parking. Building rules, etc., were established based on this basic policy that limited the maximum height to 250m, and ensured sky factors such as the elevational coverage of building to land ratios, etc. This period was as an important era in Japanese urban development history, but remarkably the conceptual and technical background to urban development in this period is not known.

Subsequently 10 high-rise buildings were built in Nishi Shinjuku, and in 1991 when the Tokyo Metropolitan Government relocated there, the new town area of SKK was completed. These high-rise buildings can be recognized as an overall group, and they have established an image for the Nishi Shinjuku new town.

Thereafter the wave of urban development proceeded to spread to the areas adjacent to the new town, due to the continuing enormous power of Shinjuku Station through which 3.5 million people pass every day. From 1975 Nihon Sekkei was involved in the re-development of Shinjuku i-Land Tower on behalf of more than 200 owners and leaseholders, which was completed in 1995 after many twists and turns.

Buildings such as Shinjuku Oak City and Nishi Shinjuku 6-chome West 6th Zone which were completed in 2002 were mainly combined office, commercial, and residential developments. The external appearance of these high-rise buildings is a continuation of the buildings of the SKK area, but these buildings have a completely different background having been created through the expenditure of much time and effort in consultations with the owners and leaseholders. The initiatives for forming the cityscape also reflect the requirements of the times, and have evolved from the initiatives for the initial special districts.

新宿智能大厦（1995）Shinjuku Minds Tower
摄影：川澄建筑摄影事务所

新宿爱岛大楼（1995）Shinjuku I-land Tower
摄影：三轮晃久摄影研究所

026	厦门海峡文化交流中心 International Convention Center, Xiamen	
034	上海浦东图书馆新馆 New Shanghai Pudong Library	
040	山东省博物馆新馆 The New Grand Shandong Museum	
044	佛山新闻中心 Foshan Media Center	
050	日本东京国立美术馆 The National Art Center, Tokyo	
052	日本长崎县美术馆 Nagasaki Prefectural Art Museum, Japan	
054	日本北海道洞爷湖国际媒体中心 Hokkaido Toyako Summit, International Media Center	
056	2010年上海世博会日本馆 The Japan Pavillion, EXPO 2010 Shanghai	

Culture and Broadcasting Media Facilities
文化媒体建筑

厦门海峡文化交流中心
International Convention Center, Xiamen

总建筑面积	127,360 ㎡
设计期间	2004/11~2006/10
建设期间	2006/11~2010/9
竣工年月	2008/9(A区)
配合设计	上海建筑设计研究院有限公司
摄影	深圳市匠力摄影设计有限公司
获奖经历	国际竞赛一等奖
	中国不动产全国房地产设计联盟年度建筑设计奖

项

目用地位于厦门吕岭路、莲前东路和环岛路的交会处。用地南侧是拥有大片绿地的国际会展中心前院。西侧和北侧的"会展北片区"为将来的开发区域，计划形成东西南北网格状城市结构。东侧遥望小金门岛。

本项目建筑轮廓线与环岛路波浪形的绿地遥相呼应，代表厦门特点的海、风、波浪、日光及空气的设计要素的运用和细致的表现，使其成为象征厦门的建筑景观。主要功能包括：国际会议中心、音乐厅、五星级度假酒店。

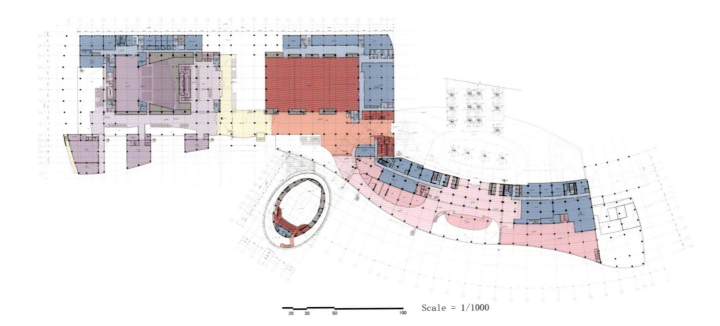

Scale = 1/1000

建筑注意体现各种节能环境技术：

1. 建筑总平面布局以及建筑设计中充分考虑对冬季日照的利用和主导风向的避开，而在夏季则利用主导风向进行建筑内部的自然通风。

2. 外墙与屋面热交换部分的室内一侧表面温度保证高于室内空气的露点温度。

3. 屋面透明部分的面积控制在屋面总体面积的20%以下。

4. 建筑内部的中庭利用自然通风实现降温的目的，不具备自然通风的区域设置了通风设备。

5. 外墙可通风面积占总体外墙面积的30%以上。其具体形式为在幕墙上设置局部开启窗或设置通风设备。

6. 建筑外部的门窗采用保温、隔热的节能措施。

7. 外窗的气密性不低于《建筑外窗气密性能分级及其检测方法》GB7107规定的4级。

8. 幕墙的气密性不低于《建筑幕墙物理性能分级》GB/T15225规定的3级。

9. 屋顶采取保温、隔热措施。保温材料采用隔热、保温性能良好的高密度聚苯乙烯发泡材料。

10. 建筑内外的墙体以轻质混凝土砌块为主，达到减轻建筑荷载、节省结构构件用量的目的。

11. 高层和裙房均采用隔热性能良好的Low-E中空玻璃，同时采用遮光百叶，达到更好的隔热效果。裙房挑空空间的上部采用电动遮光帘以及开启窗，作为隔热、通风的措施。

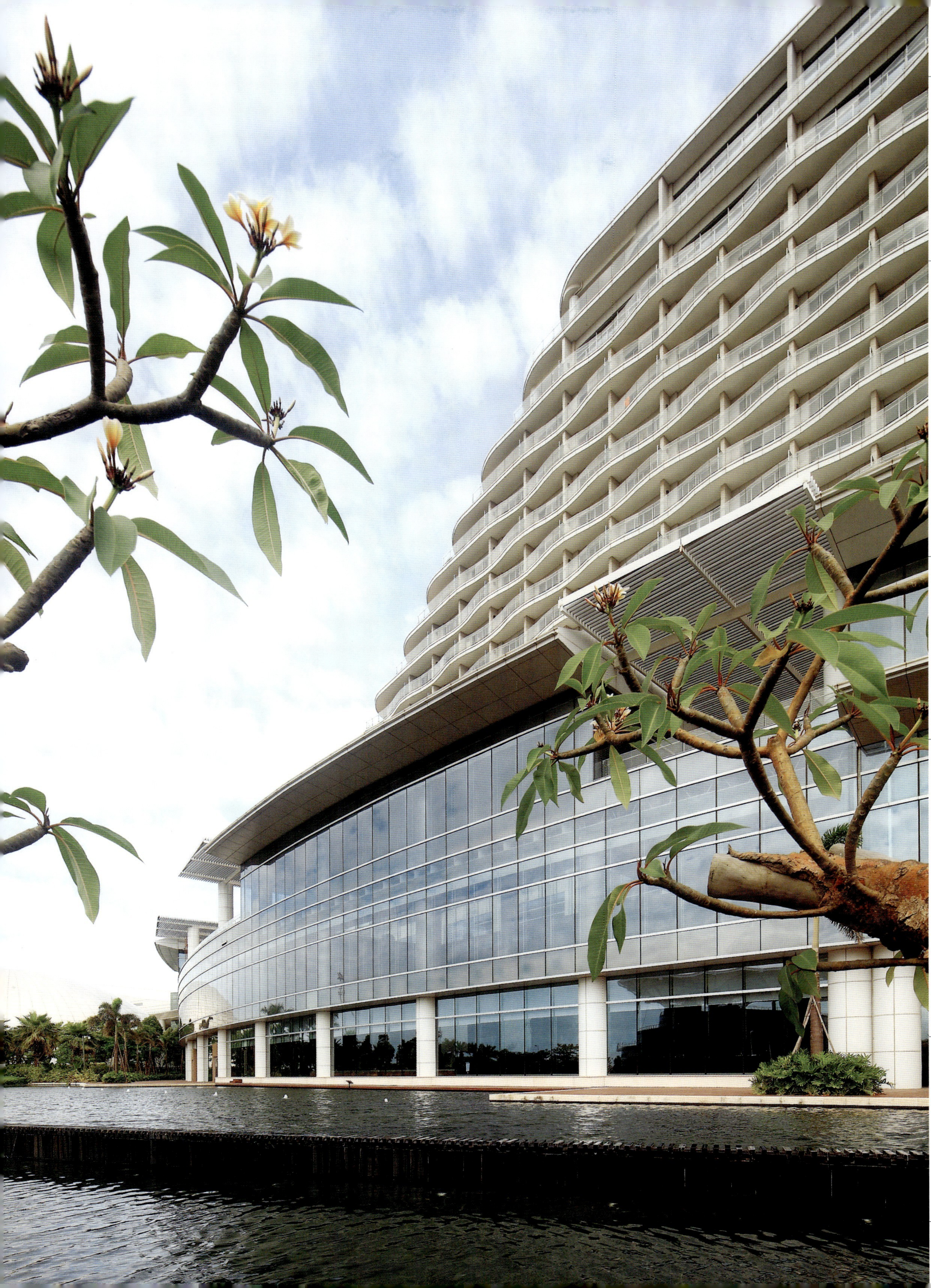

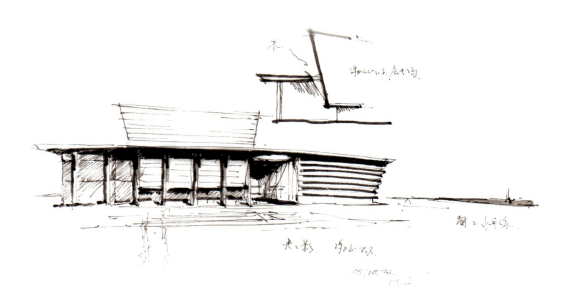

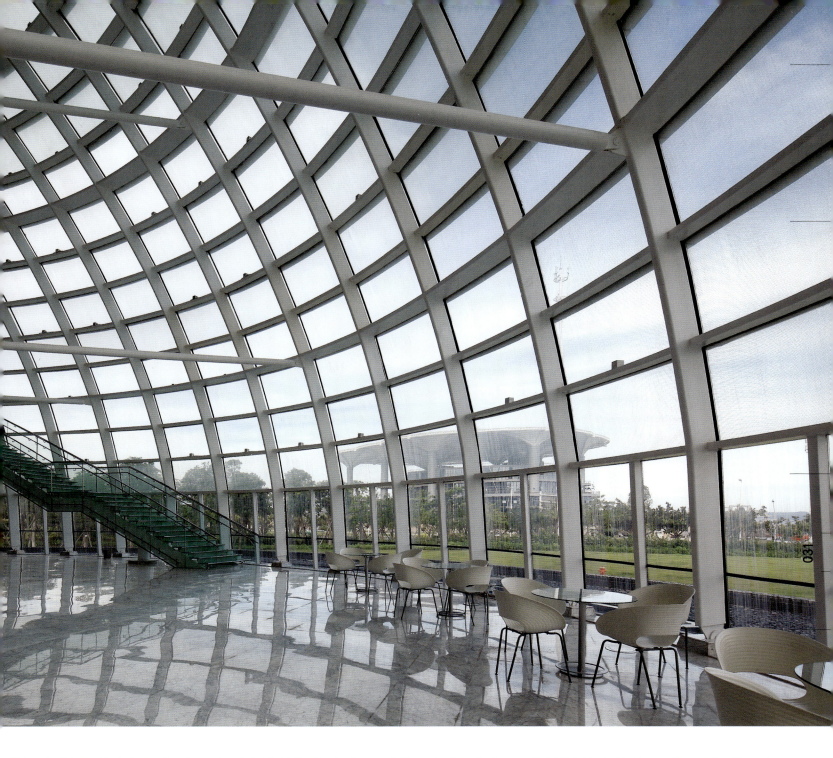

This project is located in the intersection area between Luling Road, East Lianqian Road which connect with urban areas and Island Ring Road which connects to the airport. The site faces the large planted area in the front yard of the International Convention and Exhibition Center on the south. The "North Area of Convention and Exhibition Center" to its west and north is a future development area. An east-west—south-north urban grid structure is planned for this site; the north-south axis green belt is planned while the east-west axis will be the base line for Site Planning. A distant vista of Xiaojinmen Island can be seen from the east side.

The architectural form of this project is beautiful and majestic; its rich spatial characteristics have made it a new landmark of the eastern city center in Xiamen. The construction contour line and wavy planting in the island ring road echo each other at a distance. The careful detailed approach to incorporating design elements such as sea, wind, waves, sunlight and air which represent the features of Xiamen make it a symbolic architectural landscape of Xiamen. Its main features include: International Conference Center、Concert Hall、Five-star resort hotel.

The buildings focus on utilizing a variety of energy-saving and environmental protection technologies:

1. The total construction layout and building design take full account of the use of sunshine and avoidance of dominant wind direction in winter; however, the dominant wind direction is used for natural ventilation inside the buildings in summer.

2. The design ensured that the surface temperature was higher than the dew point temperature of indoor air on the interior side of exterior walls and roof exposed to heat exchange.

3. The area of transparent part of the roof is limited to less than 20% of the total area of the roof.

4. Natural ventilation is used to achieve lower temperatures in the patio; ventilation systems are provided in the areas where natural ventilation is not possible.

5. The ventilation area of exterior wall, in the form of windows or ventilation equipment located in the curtain wall, accounts for more than 30% of the overall area of exterior wall.

6. Energy conservation measures such as thermal insulation are adopted for the external doors and windows of the building.

7. Exterior windows are specified to comply with class 4 in GB7107 or better.

8. Insulation of curtain wall is specified to comply with class 3 in GB/T15225 or better.

9. Thermal insulation measures are adopted for the roof. Insulation materials are high-density polystyrene foam with good thermal insulation and low thermal transmission characteristics.

10. The exterior and interior walls of the building are mainly light concrete blocks to reduce building load to achieve a savings in size of structural members.

11. High-rise building and podiums adopt Low-E insulating glass with good thermal insulation performance. Shading louvers are provided for further thermal insulation. The empty upper parts of the podium adopt electrically operated windows and blinds as insulation and ventilation measures.

上海浦东图书馆
New Shanghai Pudong Library

总建筑面积	60,885 ㎡
设计期间	2007/01~2008/12
建设期间	2007/10~2009/12
竣工年月	2010/06
配合设计	华东建筑设计研究院
摄影	林铭述摄影工作室以及浦东图书馆提供

本案为日本设计于2006年国际招标竞赛中获得一等奖的大型公共建筑项目，将阳光和清风引入馆内，形成一座充满智慧的图书馆建筑。

浦东图书馆基地距上海市中心仅8 km，坐落在以博览会会场和世纪公园为中心的浦东现代文化核心带上。我们的设计是在以下七个基本理念之上展开的：①个性化设计；②不增加环境负荷，保持经济活动与自然环境的平衡；③可持续发展的设计；④利用最新科技；⑤无障碍设计；⑥实现合理的投资效果；⑦从"建设"到"废弃"长远的视角、合理的规划。服务对象不问年龄、职业和国籍，建筑形态强调象征性，力求成为文化公园内醒目的"地标性"建筑。

我们设计了"绿丘",与文化公园的绿化和谐统一,在绿丘上浮现出一座外形简洁明快的方形建筑,它傍水而筑,沉稳又透着几分轻盈与俊逸。米黄色砂页岩和玻璃排列而成的富有韵律的外墙,如跃动的音阶和排列整齐的书籍一般呈现在人们的面前,也使公园南侧、未来从地铁站走出的人流和北门前道路一侧的人流的视线形成连续性。石材百叶和玻璃幕墙双层结构组成几何图案,东西向幕墙为内循环呼吸式幕墙,有助形成冬暖夏凉的室内环境。

在地下设停车场和设备机房,在地下夹层内布置书库和后勤部门,地上为公共区域,包括主要的图书阅览部门、演讲大厅等功能。地上公共区域内的"动区"、"静区(普通文献区)"、"静区(特定文献区)"又在10m内分成3层,分区明确清晰易于辨识。我们把这3层当做"3个大地",演绎出"层叠的大地景观",使图书馆建筑充满个性魅力。位于3个大地中央的高达6层的挑空中庭,气势恢弘,以两个作为景观标志的大型吊挂结构的空中花园和跨越4个楼层的南北两侧的冲孔金属屏幕为中心,演绎出简洁、明快、新颖、立体的室内中庭空间。同时,照射着经过电动百叶调控后的柔和的自然光,形成宽敞明亮的公共空间。外观简洁而有力的图书馆,内部却拥有丰富的空间构成,作为亲民的现代公共文化设施,我们力求为读者提供出与书籍交流的全新的舞台。

This project is a large public construction project designed by Nihon Sekkei which won the first prize in the 2006 International Bidding Competition; sunshine and breeze are introduced into the interior of the library; full of wisdom the library building is loved by Shanghai citizens.

The site of New Shanghai Pudong Library is only 8km from the center of Shanghai, located in the core belt of Pudong modern culture center occupying the former Expo site and the Century Park. The design is based upon the following six basic concepts: "personalized design", "sustainable design for development which does not increase environmental load, and maintains the balance between economic activities and natural environment", "adoption of the latest technology", "barrier-free design", "achieve reasonable investment returns", "rational long term planning from construction to final demolition"; " universal design for all ages, occupations and nationalities"; "symbolic architectural landmark in the cultural park".

The "Green Hill" design concept is in harmony with the surrounding planting of the cultural park, a clean simple cubic building rising on the green hummock; it is built along a river, a quiet and strong form with both lightness and prominence. Apricot cream colored sand-shale and glazing are rhythmically arranged into the exterior wall to form a vibrant musical scale. Rows of books are prominently exhibited in front; but the stream of people from south of the park are provided with a continuous line of sight through the future metro station to the road in front of the North gate. The double skin structure of stone shutters and glass curtain wall compose geometric patterns; the east-west glass curtain wall is the respiration-type glass curtain wall with integrated air recycling, which will help to create an indoor environment which is warm in winter and cool in summer.

The parking lot and equipment rooms are placed under ground, stack room and logistics sectors are also placed underground on the first floor mezzanine. Public areas are placed above ground, including the main reading rooms of library, lecture halls and other function spaces. The "activity area", "quiet area (general literature district)"and "quiet area (specific literature area)" in the public area above ground are placed on three separate floors, each 10m high corresponding to two core area floors. The partition is clear and easy to identify. These 3 floors are designed as "3 Earths", which interprets "cascade landscaping" and instills a distinctive charm to the library. The monumental atrium which is located in the center of "3 Earths" with the height of 6 floors, centered as aerial garden with large hanging structures of the 2 landscape symbols and punched metal screen four floors high at north and south sides. This interior patio space is simple, clean and innovative. A soft natural light regulated by electronically operated shutters pervades the atrium to create a bright and spacious public space.

The appearance of library is simple and powerful while its interior has rich interior spaces. As a modern public cultural facility close to the people, it strives to provide readers with a new arena for interacting with books.

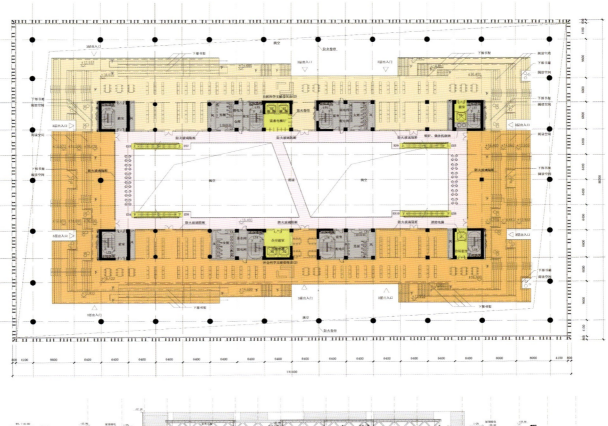
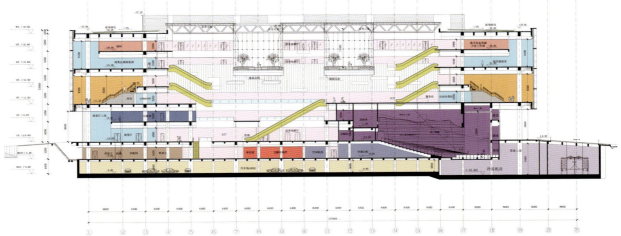

山东省博物馆新馆
The New Grand Shandong Museum

山 东省博物馆新馆是拟建的齐鲁文博中心的核心建筑，与预留的省科技馆、文博中心广场共同构成齐鲁文博中心。项目用地位于燕山立交桥以东2km处，南临经十东路，总用地面积为26.35hm²，总建筑面积为10km²，最大高差达10m。

山东省博物馆新馆的设计力图体现山东省博大精深的文化底蕴，突出山东本地文化特色，面对主入口是连接一层到三层的巨大的齐鲁文化大台阶展厅，同时也是博物馆大厅的纵向延伸，人们面对分别在不同高度的展品，宛如进入到历史发展的长廊之中，不觉地漫步到三层的齐鲁文化平台上。参观者既可以通过侧面的自动扶梯来到不同高度的主题平台，也可以步步登高，宏伟的气势让人们联想到攀登泰山，蜿蜒的布展让人们感受到黄河曲折悠长的内涵，丰富的展品令人们感叹齐鲁大地丰富的历史文化遗产。

总建筑面积	108,000㎡
设计期间	2007/10~2007/12(竞赛期间)
	2008/7~2008/3(深化期间)
建设期间	2007/12~2009/8
竣工年月	2010
配合设计	清华大学建筑设计研究院
获奖经历	国际设计竞赛优胜奖

造型设计
博物馆的造型为两个U形的紧密咬合，寓意齐文化与鲁文化的紧密联系与各自的个性特征。建筑体量在功能上分别对应展陈核心区块和公共服务区块，各成一套体系，展陈核心区块为博物馆核心功能区。建筑之间为两条带型公共空间，公共服务区块为展陈核心区块提供服务。分别为信息回廊和宣教区通廊。这些公共空间构成了两个U形的建筑体量之间的缓冲空间和联系界面，使得核心区和服务区之间形成紧密的有机联系，以一种非常理性的处理方式解决了复杂多样的各功能模块之间的相互关系。

齐鲁文化大台阶展厅是一个阳光中庭，顶部覆盖了一个轻质玻璃屋盖，屋盖悬挑出外墙面35m，形成了一个醒目的入口标志物。

齐鲁文化平台
位于三层的齐鲁文化平台是参观者一览山东省博物馆新馆内部空间的平台，宏伟的大台阶展厅一收眼底，给人们超越历史，超越现实的自豪感。

信息回廊
展厅外侧从一层到三层围绕着信息回廊，它既是展陈的内部延伸，又衬托了大台阶展厅的历史氛围。信息回廊中还分布了连接上下层的自动扶梯和休息平台。

垂直交通主要依靠4个主要核心筒和9个次要核心筒，辅以公共空间的自动扶梯和步行大台阶，为各层提供便利的垂直交通。

设计过程是一个漫长而艰辛的过程，通过和文化厅多次交流，先后提交了近10个修改方案。主创建筑师也积极参与了包括齐鲁电视台主办的电视设计辩论会，向广大市民介绍方案。今天最终建成的博物馆虽然已经和当初竞赛时完全不同，但是作为大型公共文化建筑的尝试，从建筑空间上体现当地文化，融入到体验空间中，不失为一次具有重要意义的尝试。

The New Grand Shandong Museum is the core architecture of the planned Qilu Wenbo Center. With the preserved Provincial Science and Technology Museum and Wenbo Center Square, they constitute Qilu Wenbo Centre, with a total constructed area of 100,000 square meters. The project is two kilometers east of Yanshan overpass; its south is East Jinshi Road, with total land area of 26.35 hectares; the maximum setback is 10 meters.

The design of the new Grand Shandong Museum strives to reflect the profound culture artifacts of Shandong Province, highlighting the local culture of Shandong. The main entrance is designed as the longitudinal extension to the great Qilu Culture Terrace Exhibition Hall. Visitors will encounter the exhibits at different heights, just like entering the corridor of historical development, and will naturally roam to the Qilu culture platform on the 3rd floor. Visitors can either come to theme platforms at different heights by using the side escalator or gradually climb. The magnificent upward movement will recall climbing Mount Ta, the winding exhibitions will invoke the long and meandering flow of the Yellow River and the rich exhibits will reclaim the rich historical and cultural heritage of Qilu.

Style Design

The shape of the museum is in the shape of two closely connected U shapes, symbolizing the close contact between Qi culture and Lu culture and their respective personalities. The massing scheme corresponds to functional distinction of core exhibition block and public service blocks. The public service block is an integrated system providing services for the core exhibition block, the functional core of the museum. There are three public corridor spaces between the blocks, respectively, the information, publicity and education corridors. These public spaces constitute the buffer space and contact interface between the two U-shaped construction masses, forming an intense organic link between the core area and service area. The complex relationship between diverse functional modules is solved through a very rational approach.

The Qilu Culture Terrace Exhibition Hall is an semi-open patio; its top is covered with a light glass roof with 35 meters of cantilever over the outer wall, creating a striking entrance marker for the facility. The seven functional areas of the Museum include: 1 open exhibition, 2 educational service, 3 collections storage, 4 professional scientific research, 5 security, safety and fire-fighting, 6 administration office, and 7 mechanical and electrical

The platform of Qilu culture

The platform of Qilu culture on the third floor is a platform for visitors who want to see the whole interior space of the new Grand Shandong Museum. The entire grand terrace exhibition hall can be seen; people will feel pride for going beyond history and reality.

Information gallery

Information gallery is placed around the outer exhibition hall from the 1st floor to the 3rd floor. It is not only an extension of the internal exhibition hall, but also extends the historical atmosphere of the great terrace exhibition hall. Escalators and rest platforms which connect floors are also distributed in the information corridors.

The vertical transportation mainly depends on 4 main vertical cores and 9 secondary vertical cores, with escalator and great terrace in public space as supplementary, providing convenient vertical access to the various floors.

The design process was a long and difficult process. After extended communications with the provincial cultural department, about ten amendment proposals were submitted. The chief architects have also been actively involved in the TV design debate sponsored by Qilu TV Station, and have introduced the scheme to the general public. Although the final completed museum today is completely different from that in the competition, the entire design process has great significance for large-scale public cultural buildings in realizing a building that reflects the local culture in its architectural space well integrated into the overall experience.

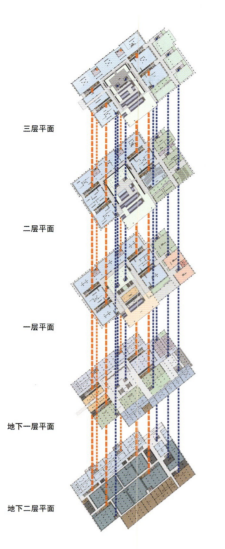

三层平面

二层平面

一层平面

地下一层平面

地下二层平面

自然与交通
展厅功能：自然馆、古代交通馆
中心展场
文博广场眺望

社会与艺术
展厅功能：文物精品馆
考古馆
民俗馆
实践功能：报告厅
模拟修复厅
电脑游艺厅
多媒体课堂
多媒体影院

历史与文化
展厅功能：山东发展史馆
石刻艺术馆
齐鲁名人馆
交流功能：贵宾室、交流展厅
入口大堂
服务功能：售票处
存包处
博物馆商店

社会参与
展示功能：城市模型展厅
实践功能：宣传教育制作
交流功能：贵宾入口
科研功能：研究室、图书资料室
文物修复室
行政办公用房
安全保卫用房

藏品与研究
库藏功能：藏品库房、临时库房
科研功能：研究室、资料室、分类室
图片制作、工作室
办公用房
设备用房

展示流线分析图

佛山新闻中心
Foshan Media Center

总建筑面积	103,000 ㎡
设计期间	2004
建设期间	2005~2007
竣工年月	2008
配合设计	深圳市立方建筑设计顾问有限公司
摄影	深圳市匠力摄影设计有限公司
获奖经历	国际设计竞赛一等奖、鲁班奖

面　向体育中心逐渐升起的连绵起伏的山丘上，高低错落的几何体建筑群围绕着多变的信息"盒子"形成大小高低的院落。一片可呼吸有表情的天幕将它们统一在一个屋檐下，人们在这里共同感受与传媒信息作伴的现代生活——这就是佛山新闻中心：山丘上的圣殿形成的景观。

佛山新闻中心与对面的体育中心分列城市景观轴大福路的两侧,形成一个既开放而又整体感强烈的城市功能空间，它将文化、生态、社会功能等诸多因素综合，借助独特的建筑形式、空间形象、环境特征和信息知觉，使之成为所在城市区域中独一无二的标志性建筑。

方网格与院落式建筑布局形式则在历史与现代的碰撞中体现人类文明传承的理念。运用节能、环保、隔热、降噪、绿化等手段，使建筑融合在自然山丘形态的开放生态系统中，阐释人与自然和谐共生的生态内涵。

独具特色的通透天幕将成为建筑的第5立面，向星空开启了一扇心窗，展示着佛山的新表情。主体办公区与服务区之间立体穿插的天桥、穿越山丘的甬道、空中漂浮的参观道与整合的信息天幕融为一体，成为新闻中心这个极具时代感的中央处理器高效性的象征。

On the rolling hills facing the sports center, the changing information "box" in surrounded by a group of geometrically shaped building of various heights, which form big and small, high and low courtyards. A piece of breathing and expression velarium unifies them in one house. Here people can experience the modem it's accompanied by media information technology, This is Foshan Media center; the it's landscape formed by the holy palace on the hill.

The Media Center and the Sports Center are situated separately on either sicn of Dafu Road the city landscape axis. Both continuously form an open and who's urban functional space, and The Media Center integrates with the factors of culture, ecology, social functions and urbanity. Distinct building types, space form, environment features and information aesthesia are integrated to create the unique landmark building in the locality.

The design integrates the building with the natural open eco-system by means of energy saving, environment protection, heat insulation, noise suppressing anc virescence to explain the ecological concept of harmonious integration of man-made natural elements. The building layout of square grid and courtyard shows the concept of human civilization heritage in the conflict between history and modernity.

The distinctive transparent velarium will become the 5th facade; the fan shaped opening is window to the sky, expressing the deepest aspirations of the New Foshan The over bridge connecting the main office area and the service area, the paved path traversing the hills, the floating visiting path are integrated whih the information velarium, which is the symbol of the Media Centher's high efficiency.

日本东京国立美术馆
The National Art Center, Tokyo

总建筑面积	49,834 ㎡
设计期间	2000/07~2002/03
建设期间	2002/07~2006/05
竣工年月	2006/05
共同设计	黑川纪章建筑都市设计事务所
摄影	SS东京株式会社
获奖经历	建筑业学会奖　GOOD DESIGN奖
	日本免震结构协会奖作品奖
	照明普及奖优秀设施奖
	日本钢结构协会业绩奖

东京国立新美术馆坐落于日本东京六本木、东京大学生产技术研究所（旧陆军第一师团步兵第三连队营房）的旧址上，其特点是馆内没有自己的收藏品，是专用于举办企划展及公募展的国立美术馆。

新国立美术馆拥有7个2000m²的展厅，总建筑面积约为5万m²，是日本规模最大的美术馆。展厅可以利用隔断分割为不同大小的展示空间，可应对各种规模的公募展及企划展，最多时可同时举办14个展览会。考虑到多个展会的同时进行，设计赋予了美术馆非常高效的搬运系统。公募展和企划展的搬运出入口完全分开，并设置了作品保管室、作品整理室、收货作业室等，空间构成充分保证了美术馆的顺利运营。

从设计之初，国立新美术馆就充分利用以青山陵园、青山公园为中心的绿地，力求打造"森林中的美术馆"，在无法确保充足的绿化用地的情况下，展开积极并富有战略性的绿化种植，使国立新美术馆的绿化最终成为连接"桧町公园—中城花园—龙土町樱花行道树—国立新美术馆—青山墓地"的区域绿化网的核心。通过一层的开放式中庭，视线穿越透明的波浪状外立面，可以将青山公园的美景纳入眼底。

日本设计建筑、结构、机电各个专业通力合作，由中空玻璃和横向百叶构成的别具特色的波浪状外立面及位于中庭极富标志性的倒圆锥形餐厅等都是经结构及机电专业设计师反复推敲与论证才得以实现的。

The National Art Center is located in the Roppongi area in Tokyo in the previous site of the Tokyo University Institute of Industrial Sciences (formerly the Imperial Army's 1st Infantry Division's 3rd regiment's barrack) as a main venue for temporary exhibitions.

The museum contains seven 2000sqm exhibition rooms, and boasts a total area of 50,000sqm, making it the largest museum in Japan. To be able to hold diverse public exhibitions and temporary exhibitions at the same time, the exhibition area can be subdivided into as many as 14 exhibition spaces. Also, to accommodate multiple exhibitions at the same time, the design also incorporates a functional moving system. The loading and unloading bays for public and temporary exhibitions are completely separated. The facility also includes a storage room for the artworks, a cataloguing room as well as stock receiving room to ensure the smooth operations of the museum.

The surrounding environment was carefully considered in the design of the building. The Aoyama cemetery and the Aoyama Park surround the museum with an abundance of greenery, giving the museum the characteristics of a "museum within a forest". In addition, many trees were planted inside of the site in the parts where greenery was not abundant. As a result, this project occupies the center of a network of green spots in the area that include the Hinokicho Park, Midtown Garden, the line of Sakura trees in Ryudocho, the National Art Center, and the Aoyama Cemetery. The open ground floor atrium opens on to a transparent curvaceous glass façade that looks out to the beautiful scenery of the Aoyama Park.

The magnificent engineering feat was achieved by repeatedly revising the design to finally attain the desired image as seen in the beautiful curvaceous façade built from paired glass and horizontal louvers, and the symbolic reversed cylinders cones on which the restaurants sit.

日本长崎县美术馆
Nagasaki Prefectural Art Museum, Japan

总建筑面积	10,225㎡
设计期间	2001/11~2002/10
建设期间	2003/04~2005/03
竣工年月	2005/03
共同设计	隈研吾建筑都市设计事务所
获奖经历	日本建筑学会作品选奖
	建筑业学会奖
	GOOD DESIGN奖
	日本建筑师会联合会奖优秀奖
	照明普及奖优秀设施奖
	日本建筑家协会奖
	长崎市都市景观奖
	预应力混凝土技术协会奖作品奖
	景观照明奖
	CS设计奖
	2005年大理石建筑奖(建筑外立面奖项第一名)
	SDA奖励奖
	国际照明设计协会优秀奖

长崎县美术馆以研究、展示明治以后的近现代美术为主，其前身是于1965年建立的长崎县立美术博物馆。美术馆坐落于常盘、出岛地区人工造地的一角，该地区作为长崎县为了实现长崎港附近的城市再生而提出的"长崎城市文艺复兴2001构想"的环节之一而被寄予厚望。

出岛运河横穿美术馆的基地中央，这一区位优势使长崎美术馆的设计极富个性及特色，使其成为一座世界罕见的跨越运河而建的美术馆。馆内沿着运河而设的美术鉴赏空间、可供来访者休息的亲水空间和立于运河之上供人穿行的空中廊桥，无不彰显着长崎县美术馆独特的设计创意。可以眺望长崎港美景的屋顶庭园与美术馆西侧海滨公园的绿化景观自然衔接，富有回游性的设计使参观者以外的人群也可以自由利用这处庭园。

长崎美术馆由两栋建筑组成，分立于运河两侧，根据使用功能划分为以展示为目的的"展览大楼"和以研究保护为目的的"美术馆大楼"。"展览大楼"配有信息检索及购物功能，并为本县居民提供美术活动场所，"美术馆大楼"内则集中设置以调查、研究、展示为目的的展厅和收藏库，形成科学合理的空间布局。利用运河之上的"空中廊桥"将这两栋建筑连接起来，而馆内面对运河而建的"光之回廊"则是可以感受四季更迭，欣赏城市景观的美术鉴赏空间。

摄影：阿野太一/FWD INC.

摄影：吾城

摄影：吾城

摄影：吾城

The Nagasaki Prefectural Art Museum was established in 1965 as a prefectural museum, and has become the major venue for post Meiji era modern art. The construction site is located in the corner of reclaimed land of Joban Dejima district, and is a part of a highly anticipated development of the Nagasaki Urban Renaissance 2001 project.

The Dejima Canal passes through the center of the project, giving the project a unique water element that makes it a memorable destination. This feature of the site is utilized to provide the visitors to the museum with a unique experience of being surrounded by canals, viewing artworks in a space at the edge of the water. The greenery of the sea side park located on the west side of the site stretches all the way to the site, and can be viewed from the rooftop garden. This space is open to the general public as well as the visitors to the museum.

The museum is characterized by 2 functional zones, the "openness" of the gallery building and the "security" of the museum building, demarcated by the canal. The gallery building is the expression of the modern museum, with devices for information search, shopping, and services to support the activities of the local residents; the museum building is the expression of the radical museum, in which attention was made to provide space for investigations, research, exhibition, and storage. The two buildings are connected by a "galleria bridge", and the sided facing the canal is called the "light cloisters" movement through which provides ever changing panoramas and sensual perceptions of the surroundings and the beautiful seasons.

日本北海道洞爷湖国际媒体中心
Hokkaido Toyako Summit, International Media Center

总建筑面积	10,693㎡
设计期间	2007/12~2008/01
建设期间	2008/01~2008/05
竣工年月	2008/05
其他设计单位	北海道开发局修建课　山下设计
环保技术支援	竹中工务店
摄影	高崎建筑摄影工房
获奖经历	GOOD DESIGN奖

集 环保技术之大成的临时建筑

北海道洞爷湖国际媒体中心是为2008年7月在北海道洞爷湖举行的西方八国首脑会议新闻报道中心而建的临时建筑。不仅拥有各国新闻媒体工作者向世界各地播报采访信息的新闻中心，还设有各国领导人召开记者招待会的会场。配合峰会的主要议题——地球环境问题，国际媒体中心在建设过程中采用了夏季冰雪制冷系统、自然采光、太阳能发电、墙面绿化等绿色环保设计手法，其中墙面绿化的植栽设计再现了北海道从海边到高山自然植被的生态景观。建筑的设计和选材以资源的循环利用为前提，国际媒体中心最终进行建筑解体时，按重量比实现了99%的资源循环利用率。

This is a temporary building constructed as a media center for the Hokkaido Summit held at Toyako. It was used by journalists from the countries of the world for filing their news articles and it was provided with rooms for press conferences with the leaders of the participating countries. As the main theme of the summit was environmental problems, environmental technologies such as snow cooling enhanced air conditioning, natural lighting, solar power generation, etc., were adopted, and the planting design for greening the wall surfaces reproduced natural vegetation from the full range of Hokkaido's seaside to mountain environments. Also, the design was carried out by selecting building materials based on the assumption that they will be used again or recycled as resources, and as a result 99% of the mass at the time of demolition was recycled or reused.

2010年上海世博会日本馆
The Japan Pavillion, EXPO 2010 Shanghai

总建筑面积	7,200 ㎡
设计期间	2008/03-2009/12
建设期间	2009/4~2010/3
竣工年月	2010/3
其他设计单位	设计总监：彦坂裕
	实施设计：竹中（中国）建设工程有限公司、
	上海现代建筑设计院（报建）
摄影	市村雄作

2010年上海世博会日本馆的设计理念为"像生命体一样会呼吸的建筑"。日本馆以减少环境负荷，并通过与自然环境进行资源、能源的互换，实现像生命体一样的建筑。外装采用被称作ETFE的新材料。建筑外皮由内藏太阳能发电板的2层ETFE膜的气枕单元构成，其外形为如生命体一样的穹顶形状。被誉为"循环呼吸柱"的6根柱子作为穹顶的结构，同时还具有采光、雨水收集、新风导入、通风、排风的功能。循环呼吸柱首先能够把自然光线引入展馆低层，实现了建筑物中央部分室内空间的自然采光。下雨时循环呼吸柱还可以收集并储存雨水。这些被储存起来的雨水可用于屋顶洒水，冷却建筑的外立面以及促进光触媒的自我清洁。洒出去的水可通过循环呼吸柱再次被收集起来，从而实现水的循环利用。同时，循环呼吸柱还能够导入新风，以此降低空调负荷，有效解决挑高展厅的排风换气问题。另外，6个循环呼吸柱中的3个，外观如同触角，富有象征意义，发挥着散热功能。其优美的造型同时也是室内设计上的一大亮点。循环呼吸柱为外观造型、室内效果、结构支撑、环保设施四位一体之集大成的建筑设计硕果。

日本馆采用了更多的环境技术、新材料，并通过外观、内部空间、结构、环境技术的统一使其成为新世纪环境建筑的样板。

This is the Project for the Japanese Government Pavilion for the 2010 Shanghai World Exposition created under the concept of "Organically Breathing Architecture". It aims to realize an architecture that reduces environmental loads by exchanging energy with the environment similar to an biological entity. The exterior is covered in an new material called ETFE. Two sheets of ETFE create a pillow structure which also contains a solar photovoltaic power unit, becoming a "Power Membrane", to create an exterior which is almost organic in appearance. A double cone shaped structure named "Eco Tube" provides the main structural support for the building, while at the same time providing the functions for introducing natural light, water and exterior air to the building. The Eco Tubes first introduce natural light into the rooms at the center of the structure by directing light into the lower levels. In rain, the Eco Tubes act as funnels to collect rainwater and lead into reservoirs. This water is used for splaying on the roof to cool and clean light catalytic self-cleaning exterior membrane. The splayed water is collected again by the Eco Tubes to be returned to the reservoirs to complete a water recycling process. The introduction and use of exterior air helps to reduce load on the air conditioning system. Furthermore, expulsion of hot air from the high ceiling exhibition spaces through the Eco Tubes to realize a natural ventilation system. Three of the five Eco Tubes are also equipped with symbolic chimney towers which are highly efficient heat dissipaters. The Eco Tubes also create extremely attractive forms in the interior design. The Eco Tubes truly unite exterior and interior design, structure and environmental building systems into a single integrated design.

Instead of adding environmental technology and new materials piecemeal into the structure, the design proposes a vision of the new generation in environmental structures, totally integrating and unifying exterior and interior design, structure and environmental building systems

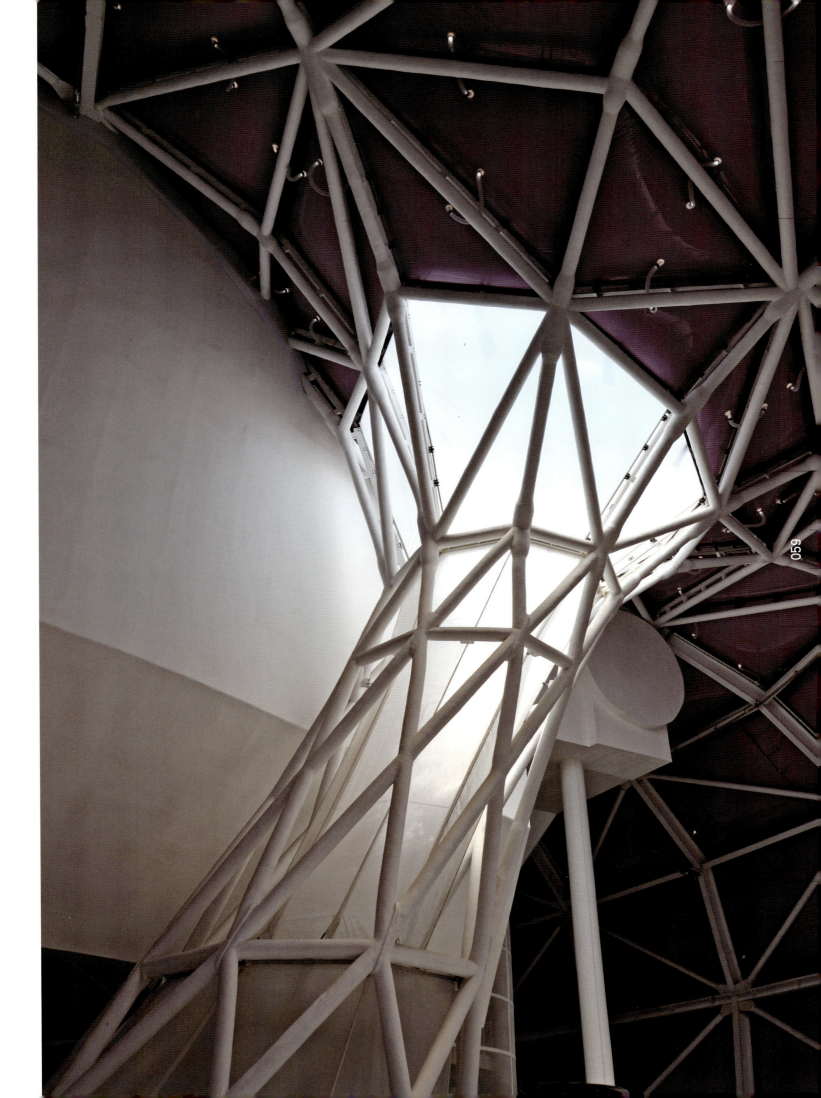

062　天津空港学校及幼儿园
Tianjin Airport Development Area Elementary School / Junior High School / Kindergarten / Day Nursery

068　日本海洋中等教育学校
Kaiyo Academy, Japan

070　日本东京大学柏研究中心
Institute of Environmental Studies, Graduate School of Frontier Sciences, The University of Tokyo

072　天津空港健身中心
Tianjing Konggang Gymnasium Center

076　日本相模女子大学雏菊馆、体育馆
Sagami Womens' University Marguerite Building / Gymnasium, Japan

Educational Facilities

学校教育建筑

天津空港学校及幼儿园
Tianjin Airport Development Area
Elementary School / Junior High School / Kindergarten / Day Nursery

总建筑面积	空港学校：19,632㎡
	幼儿园：5,498㎡
竣工年月	2011
配合设计	天津市天友建筑设计有限公司
摄影	林铭述摄影工作室
获奖经历	国际设计竞赛一等奖

本项目为2008年在天津举办的国际招标竞赛中"日本设计"被选为设计单位的学校建筑项目，是为满足当地的义务教育而建设的小学、初中一体的9年一贯制学校，相邻基地内计划建设幼儿园和托儿所。本项目利用9年一贯制学校的特点，积极引进在中国尚未普及的开放空间和组团式开放布局，旨在打造一座新颖独特的学校建筑。

9年一贯制学校
为了配合学生从小学到初中每个阶段的成长与教育，设计中将每两个年级设定为一个单元（9年级单独是一个单元），所有单元环绕位于中央的主楼呈弧形布局。随着年级的提升孩子们身边的风景、所处的空间都会发生变化，为将在这里度过9年宝贵时光的学生们带来升级的喜悦。

一个个单元犹如为孩子们度身定造的"小小家园"，其内部设有促进师生交流的公共空间和可以实现师生互动的办公室。公共空间内设有挑空和天窗，不仅构筑了内部的立体视线关系，还进一步提升了单元内的采光和通风。普通教室与开放空间之间的隔墙上设有高窗，在追求空间上

的整体效果的同时，普通教室通过采用井字密肋梁结构，与开放空间之间形成质的差异。

在设有多媒体中心、餐厅等小学生和初中生共用设施的中央主楼内，配有兼具阶梯教室和展示空间功能的专用教室，同时借助活动平台、教室组团和中庭，各年级之间的交流得到了有效的促进。并且，椭圆形设计的主楼还在视觉上确立了学校中心设施的地位。位于中央楼的两座"太阳能烟囱"（通风塔）除了发挥环保、节能的功能之外，同时作为象征环保理念的设施，可以唤起学生的环保意识。

体育馆设于半地下，与整体体量相协调。同时，无论从幼儿园还是从学校看，体育馆都是一座正面建筑。

围绕中央主楼，年级单元、体育馆呈组团型布局，形成多彩的空间和丰富的相关性，我们希望在宝贵的9年学习生活中，每个孩子都能寻找到适合自己快乐成长的空间。

幼儿园
幼儿园与学校毗邻，充分考虑基地周边的城市环境，延续学校中采用的"圆"和"单元"的设计理念，实现整个小区的和谐统一。并且与学校相同，力求为孩子们打造一座充满活力的建筑。

建筑内部按照一定的规律布置充满绿化的中庭，为孩子们提供可以亲近自然的环境，体现了关注儿童身心健康的理念。

建筑的外观凹凸有致，造型独特，富有趣味性，为建筑内部带来凹凸和高低差，形成丰富多样的室内空间。立面材质鲜亮又有对比，营造出活泼积极的氛围，符合幼儿的心理需求。

The project is a school building project by Japanese Design as a design unit in the 2008 international tender contest in Tianjin. It would be built as a 9-year school of primary school, junior high school to meet the local compulsory education. In the adjacent base, we plan to build kindergartens and nurseries. With the characteristics of 9 years consistent school, it aims to create a new unique school building by introducing open space and group-style open layout, which are not popular in China.

9 years consistent schools
In order to cater for the growth and education of students from primary to junior secondary education at each stage, each two grades will be set as a unit (grade 9 is a separate unit).All the units are around the main building in the center with curved layout. The landscape and space of students will change with the increase of grade. It will bring joy to the students who will spend 9 years precious time here.

Every single unit is just like a tailor-made "little home" for children; a public space for teacher-student communication and the office for teacher-student interaction are set inside it. The public space is designed with cantilever and skylights that not only form an internal three-dimensional view, but also further enhance the lighting and ventilation. There are high windows on the walls between common classrooms and open space; when seeking the overall effect, the common classrooms are different from the open space in quality by adopting Tic-Tac-ribbed beams.

In the central main building where there is multimedia center, restaurant and other facilities for primary and junior high school students, there is also a special classroom with terrace lecture room and exhibition space. It promotes the exchanges between grades by activity platform, classroom groups and patio. At the same time, the oval-shaped main building determines the position of school central facilities in vision. The 2 "solar chimney" (ventilation tower) located in the central building not only play a role of environmental protection and energy saving, but also will arouse the environ- mental awareness of students, as a symbol of environmental protection.

Sports Centre is located in semi-underground, coordinated with the overall body mass. At the same time, it is a positive building from kindergarten or school.

Around the main building, grade unit and Sports Centre present groups layout, forming a colorful space and rich relevance. We hope that every child can find their own space for happy growth in the 9 years school life.

Kindergarten
A kindergarten is adjacent to the school. We fully consider the surrounding environments of the base, and continue the design concept of "round" and "unit" adopted in the school, so that the whole community is in harmony. As the school, the project is aimed at building a dynamic architecture for kids.

A patio full of vegetation will be set inside the building according to certain laws, to provide kids with the environment close to nature. This reflects the concept of care about kids' physical and mental health.

The appearance of the building is convex and unique and full of interest. It brings convex and height difference for the inside of the building, thus a rich interior space is formed. The façade materials are bright and contrastive, creating a lively and positive atmosphere complying with the psychological needs of kids.

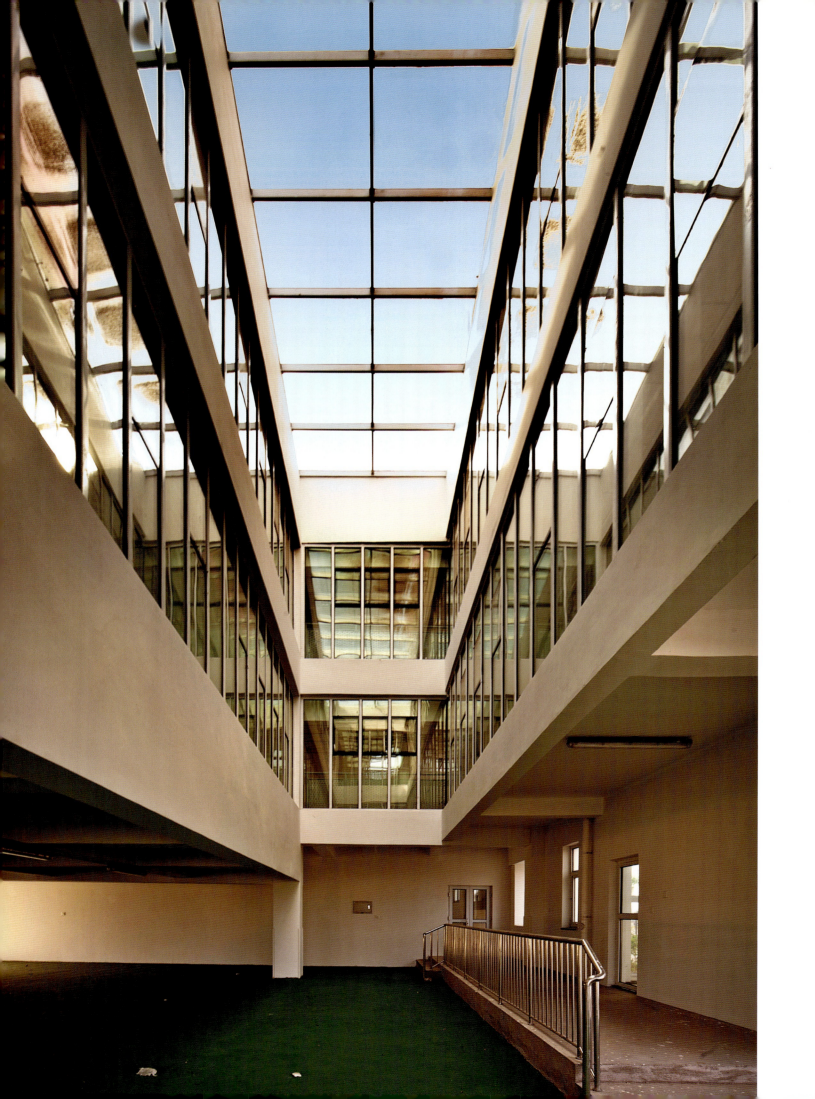

日本海洋中等教育学校
Kaiyo Academy, Japan

面 积高达13hm²的宽敞大气的海边校园，大半个基地面向平静的三河湾。为了形成培养人才的最佳环境，打造作为全寄宿制生活场所的丰富多彩的校园，体验壮美海景的建筑格局与设施规划以及自由与规则的空间构成是本案的三大设计主题。

基地的东西向长，只有西侧与道路相接，教职员宿舍靠近海边，设置从校园林荫大道通往海边球场的通道。穿过中央大楼的门厅，展现在眼前的是被教室包围的中庭形式的校园球场以及伸向大海的平台。食堂大楼和体育馆则通过草坪覆盖的小山丘与对面的立体广场相连，环抱着有小型中庭的学生宿舍群随机散落在原野上。为学生们多姿多彩的生活与学习设计相应的尺度丰富的空间，越深入校园内部，氛围愈加轻松休闲。

This is a broad 13ha campus by the sea in the calm Mikawa Bay. In order to create an environment to educate young people in the spirit of the founding of the school, there were 3 important design themes: to create a residential campus with a rich living space atmosphere, building arrangement and facility planning in which the grandeur of the sea can be experienced, and to create a space for the cultivation of freedom and discipline.

The approach was set to arrive at the front court which faces to the sea from a tree lined campus mall on the sea side of the staff accommodation. Passing through the entrance hall of the center building, you arrive at a campus court in the form of a courtyard surrounded by classrooms and a terrace that opens out to the sea. Next the dining building and the gymnasium building lead to a plaza that surrounds a grassy mound. Then you arrive at the dormitory field where a group of student dormitories are randomly arranged in a C-shape around a small courtyard. The spaces on various scales in accordance with the day to day activities of the pupils are in a sequence that becomes more casual the further you proceed.

总建筑面积	36,003 ㎡
设计期间	2003/01~2004/12
建设期间	2005/01~2006/03
竣工年月	2006/03
摄影	SS名古屋株式会社
获奖经历	日本建筑学会作品选集奖
	JIA优秀建筑选奖

日本东京大学柏研究中心
Institute of Environmental Studies, Graduate School of Frontier Sciences, The University of Tokyo

东京大学柏校园内为环境研究系而建的综合研究大楼是PFI项目之一。团队在严格遵守PFI项目所规定的详细标准的基础上，提出利用各种天然能源，以低成本实现节能的方案。例如，在中庭与楼梯间上部的玻璃盒子内设置换气扇，利用温差、压差以及风的引导效果实现换气。通过利用基坑的地热箱引入新风，促进走廊和中庭的自然通风。外墙采用将金属隔热三明治板直接固定在混凝土纯钢骨架内的施工方法，在此外墙上附设垂直与水平百叶以遮挡各个方向的日照，一年内成功削减空调负荷的36%。

Institute of Environmental Studies, the University of Tokyo locating in Kashiwa compass is one of the PFI (Private Finance Initiative) projects. This is a general research building for environmental research developed on the Kashiwa Campus as part of The University of Tokyo's first PFI project. The team proposed many methods of natural energy use and many methods to achieve low cost by energy efficiency, in strict compliance with the requirements document which was defined in detail by the PFI project. For example, ventilators were provided in the glass box at the top of the atrium and the stair well, that provide ventilation using temperature difference, pressure difference, and the wind induction effect, and natural ventilation in the corridor and atrium is promoted by linking with an external air intake from a cool tube that uses a pit. The construction of the external façade was a pure RC moment resisting skeleton frame with metal thermally insulated panels to which windows were directly fixed. By providing horizontal or vertical louvers on the external façade, used in accordance with the sunlight properties throughout the year, it was possible to reduce the annual cooling load by 36% at the planning stage.

总建筑面积	21,031 ㎡
设计期间	2003/03~2004/07
建设期间	2004/08~2005/03
竣工年月	2006/03
合作设计	大野秀敏
	东京大学大野研究室
	大成建设设计联合体
摄影	川澄建筑摄影事务所
获奖经历	NISSC ISOWAND竞赛大奖
	千叶县建筑文化奖（环保建筑领域）
	JIA环境建筑奖优秀奖（一般建筑部门）
	生态建筑奖

天津空港健身中心
Tianjin Konggang Gymnasium Center

总建筑面积	22,560 ㎡
设计期间	2009/5~2009/12
建设期间	2009/7~2010/2
竣工年月	2010/12
配合设计	天津天友建筑设计院
摄影	林铭述摄影工作室

本项目位于天津滨海新区空港加工区，是一所具有国际先进水平的健身设施场所。

主体建筑位于基地中央，基地北侧结合城市绿化带布置景观型入口广场，基地东侧和南侧布置非机动车停车位及集中室外停车场，西侧为面向城市主干道的室外运动场地，如此将室外场地分成两大区域：一是人们的活动场地；二是停车、后勤场地，做到人车分离。

建筑单体设计理念

1. 和而不同的建筑形体

   建筑简洁的四边形外轮廓使其融入区域整体规划之中，同时植入舒展流畅的曲线和曲面元素，结合精致细腻的玻璃与金属构件，又与周边建筑产生鲜明的对比，形成一道亮丽的城市风景线。

2. 空间的透明与交流、行为的互动与对话

   空港健身中心将是一座未来空港社区的重要社会交流场所，复杂的各种场馆和活动室可以隔窗相望，以实现视线四通八达，人们便能够体验到周边场馆的氛围，如此在参与不同运动项目时各自扮演着看与被看的角色。

3. 形体与空间的艺术化塑造

   生命之美在于它时时刻刻都处于运动之中，而体育，恰好集中了这种美，它将爆发与技巧、速度与优美、力量与精神同时体现到极限，赋予人无限的启示与动力，天津空港健身中心外部形体和内部空间形态正应体现现代体育的这种精神美，于是在三层的健身区域特别设置了曲面楼板，使得整个健身空间处于运动状态，而人们在高低起伏中行走，也自然成为了一种健身活动。

4. 参数化运用

   设计中运用了大量的参数化建模（rhinoscript、grasshopper），将复杂的幕墙最终简化为有规律的、适合模数化生产的构件，减少造价的同时降低了施工难度。

The project is a gym center up to advanced international level, located in the Konggang processing zone of Tianjing Binhan New District.

The main building is located in the base center. The entrance square with landscape is at the north of base. The parking area stands at the east and south of the base to serve the non-motor vehicle and central outdoor parking. At the west of base is the outdoor sports space divided into two parts, i.e. spaces respectively for the folks and parking / logistics, for the purpose of separation the folks from the vehicles.

Design Concept of Single Building
1. Harmonious Integrity with Various Style
   The simple quadrangle shape makes the building integrated into the overall planning. The sharp contrast is made between it and neighboring buildings through its smooth curve and fine glass / metal members. The single building contributes the bright scene to the city.

2. Transparent Space Facilitates the Communication
   The gym center is to become an important social communication area in the whole forthcoming Konggang Com- munity. For the easy communication, the inner of center is partitioned by the glass windows for the comprehensive sports rooms to facilitate the mutual eye contacts and experience the phenomenon from other rooms. It makes the participants watching and be watched.

3. Artistic Shape of Building
   The beauty of life lies on its ceaseless moving. The sports integrate such beauty, which demonstrates the outbreak, skill, speed, elegance, power and spirit to the limit. It endows the people with the infinite inspiration and dynamics. The design of this gym center just embodies this spirit through the curve floor slab installed in the whole three-storey building. The participants may feel they are exercising when walking on such curved floor surface.

4. Application of Parameters
   The design applies plenty of parameters to establish the modulization, e.g. rhinoscript or grasshopper. In such a way, the complicated curtain wall is simplified to the assembly of members produced through the modulization. It may reduce the cost and difficulty of construction.

日本相模女子大学雏菊馆、体育馆
Sagami Womens' University Marguerite Building/Gymnasium, Japan

总建筑面积	6,398 ㎡
设计期间	2006/06~2007/01
建设期间	2007/05~2008/03
竣工年月	2008/03
摄影	日暮摄影事务所
获奖经历	神奈川建筑竞赛优秀奖

雏菊馆静静地伫立在一片幽静的黑松林中,人们难以想像基地其实紧邻相模大野车站。它与北侧的建筑群风格迥异,是一座十分重视内部与外部空间视觉联系的建筑。为纪念建校100周年而建的雏菊馆,由设有办公室、咖啡厅、保健室、自治委员会室、花园厅等的4层大体量和设有架空空间、问讯处的一层以及设有休闲茶吧的二层小体量构成。

将混凝土墙和透明玻璃覆盖的大小不一的两个体量精心嵌入被郁郁葱葱的树木环绕的中庭空间内,设在二层的空中连廊将它们连接起来。将其作为进入小学、中学、高中的大门的象征的同时,力求为二层部分赋予生机勃勃的活力。

设有休闲茶吧的架空空间由水池、浮在水上的平台以及细柱支撑的轻盈体量构成,空间清爽而明亮。咖啡厅和休闲茶吧的内部空间被黑松、红叶、樱花、梅树等各色各样的植物环绕,位于顶层的花园厅穿过屋顶庭院与四周郁郁葱葱的树林视觉相连。

最大限度地减少树木的砍伐和移植,调整梅树林的密度和红叶树的位置,在贯穿中央的蜿蜒曲折的道路以外再修建两条小径,为整个基地赋予环游性。在面积达8000㎡的基地内设置绵延不断的地形起伏,打造外部空间的进深感和透明感,而位于基地中央的舒缓起伏的宽阔草坪,加强了地面标高与二层的视觉亲和性。

新体育馆是由两个篮球练习场组成的小型建筑。南北面上设有清水混凝土墙、绿色抽屉和门型玻璃开口,充满情趣;东西面上的镜面不锈钢板直立锁缝形成凹凸有致的反射面,将周边的建筑与绿化映照成抽象风景画。由于体育馆内都是大空间,其结构很容易形成大尺度架构。而本设计采用承载建筑两端水平力的混凝土核心筒与承载垂直荷载的钢骨材料相结合的混合结构,利用轻巧的结构材料尺寸,营造清爽明亮的内部空间。在山墙上设置门型开口,在另外两面墙上设采光井和地窗,面向室内的结构法兰盘与装修材料在同一个面上,形成简洁的盒状空间。

The Margaret Hall site is a in a wood with mainly black pine trees, so you would not think it was close to Sagami Ono Station. It was built secluded within the wood with an emphasis on the relationship between the internal space and the external space, with a different atmosphere to that of the group of buildings on the north side. Built to commemorate the 100th year anniversary of founding, Margaret Hall consists of offices, a cafeteria, a health care room, a student council meeting room, a garden hall, etc., enclosed in a 4 story large volume, and a small volume with a pilotis and information center on the first floor and a tea lounge on the second floor.

The large and small volumes consisting of concrete walls covered by transparent glass were carefully placed in the void space buried within the trees. By connecting the second floors of the 2 buildings with a light bridge, it is given the character of a gate to approach the elementary school, the junior high school, and the high school, and activities are held on the second floors.

The pilotis of the tea lounge is a brightly lit space consisting of a weightless volume floating on the pond and supported by slender columns. The internal spaces of the cafeteria and tea lounge are surrounded by various trees such as black pines, maples, cherry blossoms and Japanese apricots, etc, and the garden hall on the top floor is visually connected with these surrounding trees via the rooftop garden between them.

The density of the Japanese apricot wood and the positions of the maples, etc., were adjusted keeping felling or transplanting of trees to a minimum, and 2 small paths were constructed in addition to the winding road that passed through the center, to give connectivity to the site as a whole. The feeling of depth and transparency in the external space was brought out by superimposing gentle rises and depressions in the approximately 8,000m2 site. By creating a gentle broad grassy expanse in the center of the site, the affinity between the ground level and the second floor was strengthened.

The new gymnasium is a small building that contains 2 basketball courts for practice. The north and south sides have humorous expressions consisting of fair faced concrete walls, green drawers, and gate shaped glass openings. The #400 mirror finish stainless steel plate with vertical Japanese wax tree shingles on the east and west sides form reflective surfaces with an empty feeling, that reflects the abstracted scenery of the surrounding buildings and greenery. Gymnasiums are large spaces, so their structures tend to be large frameworks on the scale of the spaces. However, in this design the dimensions of the structural members were slender, by providing a hybrid structure in which concrete cores took the horizontal forces at both ends of the building, and the vertical forces were taken by structural steel members. The aim was for a bright, light interior space, on the gable sides there were gate type openings, and on the flat side top lights and windows extending to the ground were provided. A simple and boxy space was created by making the flanges of the structural members in the interior flush with the finishing materials.

080　南京君悦酒店
　　　Nanjing Grand Hyatt Hotel

082　长春喜来登酒店
　　　Changchun Sheraton Hotel

084　巴基斯坦某五星级酒店
　　　WTC Islamabad Hotel, Pakistan

086　日本三井大厦及东方文华酒店
　　　Nihonbashi Mitsui Tower and Mandarin Oriental Hotel, Japan

088　厦门国际航运中心
　　　Xiamen International Shipping Center

094　上海长风生态商务区7A项目
　　　Block 7A of Changfeng Ecological Business District, Shanghai

098　天津国际金融中心
　　　Tianjin International Finance Centre

102　越南某银行总部大楼项目
　　　A-Bank Head Quarter Building, Vietnam

104　日本Ao青山大厦
　　　Ao Building, Japan

106　日产先进技术开发中心
　　　Nissan Advanced Technology Center, Japan

Hotels and Offices Buildings
酒店办公建筑

南京君悦酒店
Nanjing Grand Hyatt Hotel

总建筑面积	66,000 ㎡
设计期间	2003/8~2004/3
建设期间	2005/11~2008/7
竣工年月	2008年8月
配合设计	华东建筑设计院(上海)
室内设计	Bilkey Llinas Design
厨房设计	CKP

这座高级城市酒店位于中国古都南京，毗邻长江。整体规划将建筑的主要立面向凝聚了中国历史的长江展开：一条优美的弧线勾画出江岸的同时，客房沿着走廊单面展开，确保了所有的客房都具有良好的观景效果，弥补了单面布置所带来的低效率。同时，为了满足客人欣赏江景的需求，酒店的310间客房都布置了具有较大进深的阳台。

This luxury hotel is located in the ancient Chinese capital of Nanjing, situated on the banks of the Yangtze River. The overall planning emphasizes the building's broad main facade towards the Yangtze River which is rich in Chinese history: while an elegant arc line draws the outline of the River bank, the guest rooms spread in a single bank along the corridor, which guarantees all guest rooms have an excellent time, all guest rooms have larger balconies in order to enhance the enjoyment of the rive views.

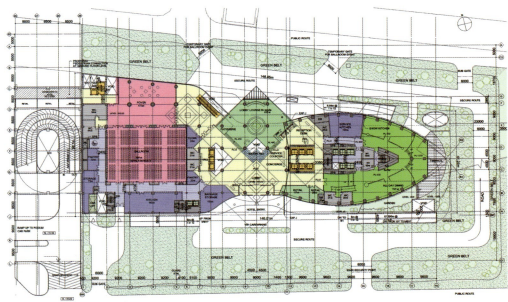

长春喜来登酒店
Changchun Sheraton Hotel

总建筑面积	156,495 ㎡
设计期间	2007/5~2008/8

本项目位于长春净月潭旅游经济开发区。这里定位为环境优美的城市生态旅游区、科技文化区、高教区和高档住宅区。作为定位为"国际上最具影响力的五星级休闲度假酒店"，用地内被赋予了以旅游休闲度假产业为核心，涵盖了五星级度假酒店，国际会议传媒中心，酒店公寓，度假公寓及休闲娱乐配套等功能。

首先将五大功能分为两个组团。会议传媒设施和休闲娱乐配套设施相对人流量较大，公共性更强，分为一组。五星级酒店，酒店公寓，度假公寓均属居住设施，到访者的目的性强，对安静环境以及私密性要求更高，分为另一组。结合用地的特殊地理位置和形状，将公共组团和私密组团分设在用地的西侧和东侧。在分组团的基础上，结合用地内的景观设计，将各功能反映在用地的不同位置。

项目以酒店为中心，围绕"山景""水景""林景"三大主体景观将生态建筑的设计理念延伸至景观设计的每个角落。

(1) 青山：会议传媒中心和休闲娱乐配套设施围绕的空间做成鲜花盛开，起伏有致的开阔山丘，通往会议传媒中心的路径从旁边经过，即使流线较长也不会感到单调。

(2) 秀水：确保用地东侧的大面积水面，度假公寓均临水而建，与自然景观浑然一体，形成绝佳的居住环境的同时，也构成了酒店和酒店公寓视觉景观。

(3) 茂林：在五星级酒店的主入口与道路之间，集中设置大面积树林，将丁二十二路的尘器隔绝在外。入住者通过自然形成的林间漫道到达酒店入口，尽管离开市政道路并不是很远，却仿佛进入了森林深处。

The project is located in Jingyuetan Tourism Economic Development Area of Changchun. This area is designated as a beautiful urban eco-tourism area, scientific and cultural areas, higher education area and high-end residential area. Given its position.As "the most influential and international five-star resort hotel", the project is the core facility for the tourism and leisure industry, covering the functions of five-star resort hotel, international conference media center, hotel apartments, holiday apartments and supporting facilities for leisure and entertainment.

Firstly, the five functions are classified into two groups. The conference media facilities and commercial supporting facilities for entertainment are grouped together as they will have relatively larger visitor flow and media publicity events. Five-star hotels, hotel apartments, holiday apartments are all residential facilities are also grouped together since their visitors have strong purpose with high requirements for quiet and privacy. Based on the geographical characteristics and shape of the site, public group and private group are separated into the west and east. The landscape design reflects the various functions of each sub-group in different positions on site.

The three main ecological design concepts of "mountain", "water" "woods" extend to every corner of the landscape design, centering on the hotel, the main facility of the project.

(1) The magnificent mountains, beautiful river, and lush trees of Jingyuetan National Forest Park have created a very special environment. This is an attractive tourism resource for urban residents.

(2) After refinement, these elements are recreated in miniature on the site, organically integrated with the main constructions.

(3) The blend of architecture and landscape is also the combination of human and natural environment that realizes the harmonious coexistence of humanity and nature. At the same time, the hotel rooms with the herringbone layout look out on different types of natural landscapes and different visual effects formed by the landscape, offering guests a personalized landscape vista.

巴基斯坦某五星级酒店
WTC Islamabad Hotel, Pakistan

总建筑面积	106,931 ㎡
设计期间	2006/8~2007/11
配合设计	ABDUL RAZZAK ASSOCIATES
	ALLIANCE CONSULTANTS
	SEM ENGINEERS
	ELEKEN ASSOCIATES

酒店坐落在伊斯兰堡郊区，周边地区属于国防部房屋管理局，拥有444间客房规模的五星级酒店。

基地南北边界水平高差为6m。为了充分利用基地独特地形的优势，将建筑分为两个部分。在酒店的两个楼层分别设置两个不同的入口。在地面标高设置南侧酒店入口，在更低处，设置北侧会议厅入口。酒店塔楼位于大堂茶座的东面，而裙房内的主要会议设施位于西面。建筑分开布局便于分散人流，使进出更为便捷。

酒店大厦高达147.28m，共有3个设备楼层，分别位于四层、十八层和三十一层，444间不同形式的客房分布在26个楼层，其中包括位于十九层和三十层的行政客房。

外立面设计采用几何形状，曲线优美，部分光滑，部分粗糙，层次分明。

The surrounding area belongs to and will be developed by the Defense Housing Authority (DHA). Rixos proposes to operate the Hotel as a five star operation with approximately 444 rooms.

The site has a level difference of approximately 6m between the north and south boundaries. The building is divided into two principal parts to take advantage of the site's unique topographical features. The Hotel is designed with two principle entrances at different floor levels. At Ground Floor level, to the south of the Lobby Lounge is the Hotel Entrance. At Lower Ground level, to the north of the Lobby Lounge, is the Conference Entrance. To the east of the Lobby Lounge lies the Hotel Tower, to the west are located the majority of Conference facilities in the Podium. The separation promotes ease of access and allows discrete separation of user groups.

The Hotel Tower, including 3 Mechanical Floors at 4th, 18th and 31st Floor levels, is 147.280m high from Ground Floor level and contains 444 rooms on 26 floors in a flexible and varied mix of rooms and suites, with executive facilities on the 19th and 30th Floor levels. The Hotel is equipped to meet the full range of their guest's accommodation and service needs.

The facades are designed on a radial geometry to offer a soft, curvaceous, part glazed, part solid, layered elevation to the surroundings.

日本三井大厦及东方文华酒店
Nihonbashi Mitsui Tower and Mandarin Oriental Hotel, Japan

日本桥三井大厦——保留与开发的完美结合

旧三井大厦是三井财阀的办公大楼，具有本部功能，于1929年竣工，拥有花岗石外檐、大理石装饰面、多立克柱式内部大厅，是美国布杂学院式近代写字楼，因其完美的设计和建筑历史文化价值，于1998年12月被指定为国家重要文化遗产。

日本三井大厦的设计为了确保与旧三井大厦的和谐，严格遵守了特定街区城市规划的规定，秉承旧三井大厦的古典风格，采用对称，建筑正面明确的平面形态，建筑的檐口线脚与旧三井大厦持平，形成并排的立面布局，结合材料的连续性，与旧三井大厦形成统一的风格。城市设计的规定还体现在建筑设计中。在建筑功能方面日本三井大厦实现了可灵活变化、易于维修管理的空间划分，同时在结构方面采用可对应1.5倍平常地震强度的3级抗震设计，设备方面则采用降低环境负荷的冰蓄冷、大温差空调、自动控制照明等环保设计手法，使整栋大楼成为各项功能完善的现代化建筑。如果说旧三井大厦是继承传统的象征，与其并排的日本三井大厦则是展示未来方向的区域象征。

Conservation and Development of Mitsui Main Building

Mitsui Main Building was completed in 1929 and functioned as the headquarters of the Mitsui Zaibatsu (conglomerate). With a granite façade and an internal hall with Dorian columns finished with marble, it was a modern office building in the American Beaux-Arts style. In December 1998 it was designated as a national important cultural property in recognition of the high level of architectural design and high quality construction.

In order to ensure harmony with the Mitsui Main Building, a strict code was established within the special district urban plan for the design of the tower. This code included continuation of the traditional elements of the Main Building; plan shape having a symmetrical and clear frontage, elevation set back behind the cornice line of the Main Building, creation of a streetscape with a feeling of unity with the Main Building, etc., and including continuity in use of materials, etc. These urban design codes were passed on and materialized in the architectural design. The tower function included allocation of space to accommodate changes and the requirements of maintenance and management, level 3 seismic resistance postulating earthquake forces 1.5 times higher than normal, ice thermal storage to reduce the environmental loads, large temperature difference air conditioning, automatically controlled lighting, and other detailed services design. If the Mitsui Main Building is a symbol of the desire to carry on the past, the Nihonbashi Mitsui Tower is a symbol of the future direction of the district.

总建筑面积	194,309 ㎡
设计期间	2000/07~2002/04
建设期间	2002/05~2005/07
竣工年月	2005/07
合作设计	Pelli Clarke Pelli Architects Inc.
摄影	川澄建筑摄影事务所
获奖经历	GOOD DESIGN奖、日本建筑学会业绩奖、建筑业学会奖奖

厦门国际航运中心
Xiamen International Shipping Center

总建筑面积	156,000 ㎡
设计期间	2006/11~2008/11
建设期间	2008/11~2010
竣工年月	2011/6
配合设计	厦门市建筑设计院
摄影	深圳市匠力摄影设计有限公司
获奖经历	国际设计竞赛一等奖

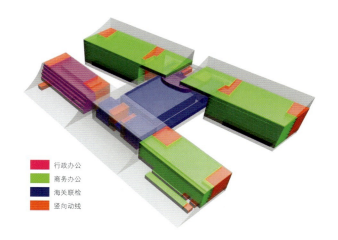

行政办公
商务办公
海关联检
竖向动线

项目基地位于距厦门岛中心西北8.5km的象屿保税区，距离海岸线仅550m，整片基地视野极其开阔，与海沧大桥遥遥相望。同时厦门机场的空中航线经过基地上空，从飞机上可以清楚地俯视整个建筑群的全貌。得天独厚的地理位置使厦门国际航运中心成为代表厦门形象的地标性建筑。

建筑的外形设计体现了现代与传统的融合。院落式整体布局和连续水平线条形成了简洁舒展而又气势磅礴的建筑形体，最大限度地与海天一线的基地大环境相呼应。而在建筑细部处理上则采用了隐喻手法，体现出了对闽南民居的院落、重檐屋顶、门廊、骑楼、花窗等传统元素的借鉴，同时也非常吻合厦门温暖多雨气候的生活方式。建筑材料方面大胆采用了闽南民居常见的红砖和花岗石，与轻巧的玻璃幕墙形成色彩与材质的对比。

航运中心共由5栋建筑组成。行政楼和办公楼围绕海关联检中心形成院落式布局，使建筑既保持了各自的独立性，同时又维系功能上的相互关联。各栋建筑均设有独立的回车场地与门厅。建筑之间由半室外空间回廊连接，即使在雨天，也可以方便通畅地来往于各栋大楼之间。整个建筑群南低北高，各栋建筑沿东西轴布局，在保证采光的同时最大限度地降低了日照辐射带来的建筑表面热负荷。

厦门国际航运中心将海关、办公、行政三项功能有机地融合为一体，运用现代的建筑材料和技术重新诠释传统的地方特色建筑文化，体现了国际化精神与厦门地域特色的交融。

The project base is located in the Xiangyu Bonded Trade Area 8.5 kilometers northwest from the center of Xiamen Island, and only 550 meters from the coast. The entire bonded area can be seen from the Haicang Bridge. Also, the flight path into Xiamen Airport lies over the logistics base. The entire layout of the buildings can be clearly seen from aircraft. The superior geographical location makes the Xiamen International Logistics Center a landmark representing Xiamen.

The design of the building reflects the fusion of modernity and tradition. The overall courtyard-like layout and continuous vertical lines form a simple, magnificent architectural form, echoing the boundless sea and sky. Metaphor is used in the treatment of architectural details, drawing from the courtyard, roof eaves, porch, veranda, windows and other traditional elements of Southern Fujian. It also corresponds to the Xiamen lifestyle with its warm and rainy climate. Red brick and granite which are popular traditional materials in Southern Fujian are boldly contrasted with light glass curtain wall in color and texture.

Xiamen International Logistics Center is composed of five buildings. The administrative building and office buildings around the Joint Customs Inspection Center form a courtyard-like layout; allowing the building to maintain their physical independence while maintaining functional interdependence. All the buildings are designed with independent car space and entrance hall. The buildings are connected by the semi-open corridors to allow easy move between all the buildings even on rainy days. All buildings are low in height in north and high in south and laid out along east-west axis. All buildings are ensured to receive sunlight, the solar heat load on building surfaces are reduced to the minimum.

Xiamen International Logistics Center organically integrates the three functions of customs, offices and administration into one. By using modern building materials and technologies, it re-interprets the traditional local architectural culture, reflecting the blend of the international spirit and Xiamen geographical characteristics.

建筑形态的推敲过程

1

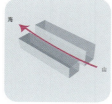
2

3

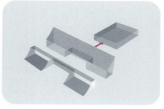
4

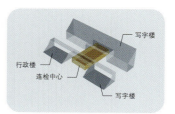
5

上海长风生态商务区7A项目
Block 7A of Changfeng Ecological Business District, Shanghai

总建筑面积	112,680㎡(北区)46,953㎡(南区)
设计期间	2006/8~2010/8
建设期间	2009/1~
竣工年月	2012/6
配合设计	上海建筑设计研究院有限公司
获奖经历	国际设计竞赛一等奖

长风地区位于上海市中心的西部,内外环之间,是上海市著名景观之一苏州河沿岸的重要开发区。本案规划用地面向中环线与主干道金沙江路的交叉点,横跨两个街区的本规划用地,位于长风生态商务区的西北角,因此未来这里将成为从西北方向进入长风生态商务区的门户。

本地块紧邻中环线、交通干道、地铁车站以及两条河流,是极具潜力和开发价值的地块,通过采用全新的设计概念,创造出引领时代潮流,预示未来发展模式的地下空间,建设地上设施与地下空间融为一体,充满活力的城市空间。

北区为高级创意办公楼,采用双塔形式,地上15层,建筑高度70m。南区的高级创意办公大楼,地上11层,建筑高度63m。

金沙江路下部规划有地铁13号线,在地下二层标高连接南北区的商业设施。

Changfeng area is located west of central Shanghai between the inner ring and outer ring roads, and is an important development area along Suzhou River, which is one of Shanghai's famous landscapes. The project site faces the intersection of the middle ring road and the main Jinsha River Road and occupies two blocks located in the northwest corner of Changfeng Ecological Business District. The site is planned to become the future northwest portal to the Changfeng Ecological Business District

The site is adjacent to the middle ring road, highway, metro stations and two rivers, and has great potential and development value. Using a new design concept, a trendsetting underground space that provides a model for future development was created. The above ground and underground spaces are integrated into a single, fully vital structure.

A twin-tower high-end office building is located in the north part, with 15 floors on the ground and a height of 70 meters. A creative industry office building is in the south, with 11 floors and a height of 63 meters.

In the lower part of Jinshajiang Road, subway line no.15 is planned. The station is connected to the commercial facilities in the 2nd basement floor of the north and south area.

天津国际金融中心
Tianjin International Finance Centre

总建筑面积	82,311㎡
设计期间	2005/7~2006/3
建设期间	2006/10~2009/10
竣工年月	2010/10
配合设计	天津市设计院
摄影	林铭述摄影工作室

本工程为友谊路国际金融中心项目，该工程位于天津市河西区友谊路与增进道交叉口处。规划可用地东西向长70m，南北向长160m，总用地面积11090m²。该用地西邻友谊路，北靠北方金融大厦，东侧为住宅小区，南邻增进道。本工程位于天津市中心，周边市政设施完善，环境良好，交通便利。

本工程整体沿南北向呈"一"字型，南侧为主楼，高166m，北侧为5层裙楼，高24m；地下两层为设备用房及机械式机动车停车库。由于本工程用地较紧张，在满足规划消防和日照遮挡间距的前提下合理规划建筑布局及结合周围道路情况设计交通组织及环境绿化设计，在裙楼五层屋顶上进行绿化设计，使环境更加灵活和人性化。

This project is an international financial center project on Friendship Road, located at the intersections between Friendship Road and Zengjin Road in Hexi District of Tianjin. The available site is 70 meters long from east to west, 160 meters long from north to south, with a total area of 11,090 square meters. This land is faces the Friendship Road on the west, with the North Finance Building to the north, the residential area to the east, and the Zengjin Road on the south. The project is located in the center of Tianjin; it is located adjoining good municipal facilities with beautiful environment and convenient transportation.

This project is designed as a Line-shape. The main building with a height of 166 meters is on the south and the 5-floor podium with height of 24 meters is on the north. The two underground floors are equipment spaces and parking garage for motor vehicles. The site area is limited and required rational design to meet requirements for fire-fighting and sun block distance in combination with traffic organization and environmental green including the surrounding roads. Planting was designed on the 5th floor roof of the podium to make the environment more flexible and humane.

越南某银行总部大楼项目
A-Bank Head Quarter Building, Vietnam

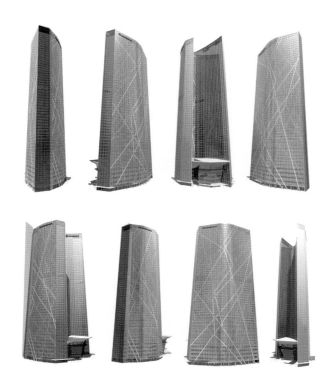

总建筑面积	173,167 ㎡
设计期间	2008/10~2008/12

本项目为越南河内市新市中心的40层高的银行总部大厦。该项目所在地汇聚着高度超过70层的川南项目等超大型项目。由于该区域为大河内规划的中枢,将会吸引越南的大型企业进驻。A银行作为长期以来深深扎根于有关农业的银行,在越南的知名度非常高。本项目的建筑设计以"农业"和与其有关的"成长"作为主题。

涌浪塔
银行致力于投资农业,发放中长期贷款,支持农村基础设施建设,发展农林牧副渔,这非常有助于农业的产业化及现代化。"涌浪塔"即标志着银行对农业和农村发展的贡献,并力求扩大其影响。

建筑中融入自然元素代表农业的发展。通过彩釉玻璃立面映照出作为雨水之源的"天空和云彩",立面上的蓝色照明代表促进生长的能量之源"雨水",会议大厅代表了收集雨水,为种子提供发芽和生长的温床的"大地",立面上的黄色照明代表了在各种能量催化之下的种子的"发芽"。"涌浪塔"象征着农业中的"生长与发展"的理念。建筑设计采用尖端科技,打造代表着创新与权利的A-BANK建筑。

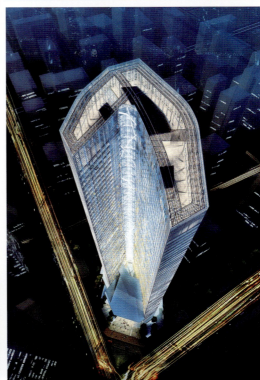

This 40-story bank headquarters building is a new city center of Hanoi, Vietnam. The bank, as agriculture has long been deeply rooted in the developing, is well known in Vietnam. The main concept of the desing is about "Agriculture" and "growth".

"Surging Tower"
A-BANK focuses on investment for rural development using medium and long-term credit to build up infrastructures for agriculture, forestry and fishery. It is a leading proponent of rural and agricultural industrialization and modernization. The iconic "Surging Tower" identifies A-BANK's powerful contribution to agricultural and rural development and willingness to expand its ambition.

The building translates the agricultural cycle of appearance and composition of elements into a landmark design. "Sky and Clouds" are the natural source for rain which is reflected in the ceramic printed façade design. "Rain" is the energy source to promote the process of growth represented by a blue lighting duct on the façade. "Field" is the ground that collects water for providing soil, or enable growth or germination to produce new plant is represented in the Convention Hall. As a result, the seed commences to germinate with all the energy sources combined and this "Germination" is represented by the yellow lighting duct on the façade. The whole agricultural cycle generates the idea of "Growth and Development" which is symbolized in the "Surging Tower." As a representation of the innovation and power of A-BANK, the design of the building expresses high-tech development with the most modern and contemporary design and construction technology.

日本Ao青山大厦
Ao Building, Japan

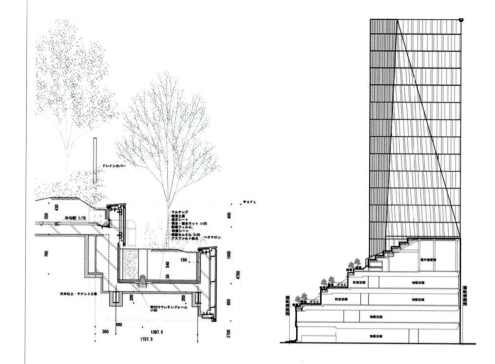

Ao 青山大厦坐落于日本时尚潮流中心青山表参道，是以餐饮、商品销售、日本服务为主的商业用途的租赁型大楼。向空中扩展的下窄上宽的楔形塔楼，绿意盎然的裙房阶梯花园，独特的外观造型使 Ao 大厦一跃成为城市的新地标性景观。

Ao 大厦的设计重点是确保建筑的商业用途，同时降低对周边建筑的日照影响及楼间风等环境方面影响的结果。Ao 大厦的建筑形状与布局是综合考虑上述要素，通过精确的电脑模拟演示最终确定的。

为了确保可以俯视东京全貌的眺望视线和表现每个租户的特色，Ao 大厦的外立面采用了全面双层玻璃幕墙（内循环式），同时利用富有韵律感的幕墙分隔及与之相呼应的不同可视光透过率的幕墙玻璃，照度各异的高辉度全色 LED 照明形成具有视觉冲击力的外立面。

位于裙房部分的阶梯花园由可供租户使用的绿色平台及亲水平台构成，在这里可以尽情欣赏由代代木公园和附近超高层大楼组成的富有东京特色的全景立体画。

在一层沿基地 3 面道路打造的 1.5m 宽、散步道状的庭园不仅景观优美，还易于亲近，为每个租户提供了一处可尽情嬉戏的室外活动空间。

人们还可以在悬浮于空中的室外阶梯上自由漫步，作为可以感受四季流转的商业空间，Ao 大厦为青山表参道增添了一个崭新的充满绿荫与自然情趣的城市新景点。

Ao building, located in the bustling and fashionable Aoyama Omotesando district, is a tenant building for restaurants, shopping and services. The shape of the building seems to expand as it stretches into the sky, and on the opposite side, a stepped garden enticingly opens to the street level. The unique shape of the building makes it an outstanding new landmark in this trendy district.

One of the prime conditions of the project was maximizing return on investment. The shape of the building which expresses a larger top and bottom with a narrow middle part is the result of certain considerations such as rentable area ratio, where the advantage at the top is the better views and at the bottom is ease of access. The design maximizes the area with higher rental revenue potential. Environmental considerations, such as simulation studies for sun shadow control and wind turbulence, also shaped the design and the shape of the building.

The outer cladding was designed so that all sides are open to views to the outside and also to fully allow the tenants to be visible from the outside. The façade uses full height glass double-skin curtain wall (air flow window type) design for maximum visibility. As a tenant building housing an abundance of diverse functions and businesses, the problem for the designers was to maintain unity in the design. The designers decided to use different layers of random patterns of graphics by playing with the rhythmical patterns of the window sashes, the transparencies of the types of glasses, and the luminance of the full color LED lights to attenuate the disorder.

The step garden consists of the green terrace and the water terrace which can be used by the tenants. This terrace boasts fabulous views towards Yoyogi Park and the skyscrapers of Shinjuku. Also, on 3 sides of the site at ground level is a garden with a 1.5m wide walkway. This allows all the tenants to take advantage of and actually use the exterior spaces.

A walk on the floating exterior staircase allows one to enjoy the diversity of the shops as well as changing colors of the seasons. This project is welcome addition to the pleasures of the life in the Aoyama Omotesando area.

总建筑面积	21,932㎡
设计期间	2001/10~2006/06
建设期间	2006/10~2008/10
竣工年月	2008/10
摄影师	木田胜久/FOTOTECA
获奖经历	国际照明设计协会优秀奖、照明普及奖

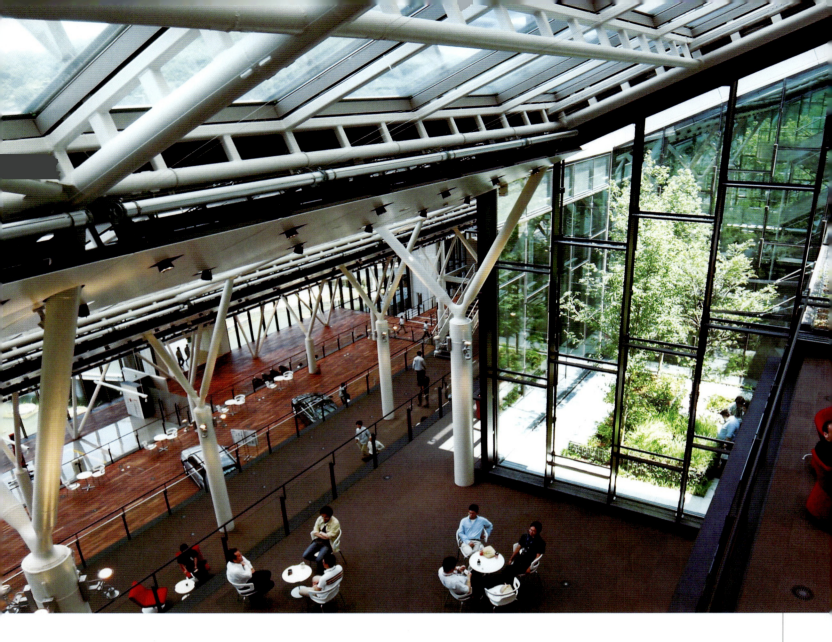

日产先进技术开发中心
NISSAN Advanced Technology Center, Japan

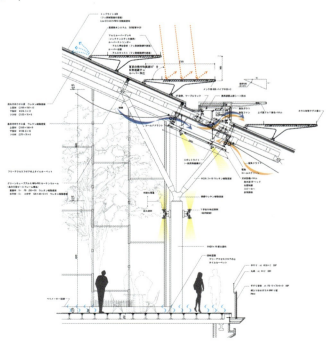

总建筑面积	69,471 ㎡
设计期间	2004/01~2004/11
建设期间	2005/06~2007/05
竣工年月	2007/05
摄影	木田胜久/FOTOTECA, 日本設計
获奖经历	GOOD DESIGN奖、建筑业学会奖、
	JIA环境建筑奖、东京建筑奖、生态建筑奖、
	神奈川县建筑竞赛奖、
	日经新办公楼奖（经济产业大臣奖）、
	免震结构协会奖

基 地东北面有高低起伏的爱名绿地、西接新城住宅区，住宅区的背后则是七沢森林公园和丹沢山，自然景观极其丰富。针对日产汽车提出的"创造／多功能／交流"的理念，如何有效利用周围丰富的自然资源成为了决定建筑形态的重要元素。对于促进创造性活动的空间，虽无确切的定义，但这一空间绝对不是安静的、毫无刺激的日常作业空间，而应该是激发大脑意识活动，激活创新思维的刺激性空间。同时，为了积极获取大自然赋予的各种刺激、大幅增加沟通及交流的空间，沿绿丘坡面打造阶梯状办公空间，上部覆盖采光顶，形成阶梯工作室。

为了融入周边树林的生态系统之中，最初设在玻璃采光顶内部的植栽，通过取消植栽上部的玻璃顶，被置于室外。同时为了能够将大自然赋予的各种感官刺激有效引入到办公空间内部，整栋建筑被塑造成为一个四面被玻璃幕墙环绕的绿色立方体。

作为技术开发中心，其办公空间应该能够灵活变化，以对应各种组织形态的多功能团队的运行需求，因此在基地空间允许的最大范围内，打造出工作大平台。并通过垂直动线及空调设备空间的通透化设计，免震结构，轻快的采光顶结构和细窗框，大跨距和纤细的柱子，实现了最佳景观视线及视线畅通的大办公空间。

The site of the project is surrounded by an abundance of natural landscapes, such as the greenery of the Aina hill in the north east, the verdant new town residential area to the west with the Nanasawa Forest Park in the background and Tanzawa mountains in the distance. A design theme of the project was to maximally integrate the surrounding natural abundance, while responding to the concepts given by Nissan; "Creative", "Cross-Functional" and "Communication". These concepts are what mainly give shape to the architectural form of the building. It is impossible to define a space that expresses Creativity. However, it should be a space that excites the senses, inspires and promotes intellectual thinking. It should not be a mundane, boring space commonly associated with routine work.

The design also encourages more communication by engaging nature into the design as an instigator of conversation. The design is in a stepped configuration facing the greenery of the hill and covered by a large skylight roof thus giving it its name as the "Step Workplace". Also the greenery planted in the inside of the skylight roof is a continuation of the natural elements of the surrounding forest. To introduce the natural elements into the office space, the four sides of the buildings are transparent curtain walls, creating a giant glass cube. Since the organization of the Institute is cross functional in structure, the spaces also need to respond to its needs in a flexible manner. The floor plate of the office was expanded to the limit, and the circulation and ventilation equipments located within the skeleton of the building, giving the building a sense of openness, and visibility.

A base isolation system to protect against earthquakes is employed that allows light structural frames with large spans and thin columns and also lighter members for the skylight and curtain wall creating a large space with great transparency and good views.

摄影：日本設計

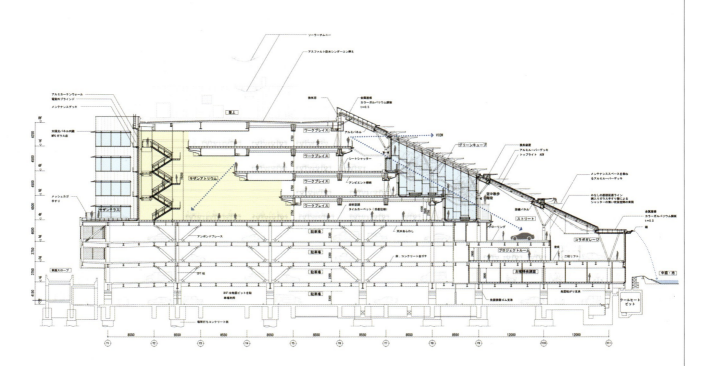

断面图 S=1:300

112	沈阳利福商业项目 Life Style Shopping Mall, Shenyang	
118	深圳宝安中心区N7综合体 Project N7, Central Business Zone, Bao'an, Shenzhen	
124	中航昆山城 Zhonghang Kunshan City	
130	威海宝泉中心（威高广场） Weihai Baoquan Complex (Wei Gao Plaza)	
136	日本北九州河岸广场 River Walk Kitakyushu, Japan	
138	日本桥一丁目大厦 Nihonbashi 1-chome Building (COREDO), Japan	

Commercial Facilities
商业建筑

沈阳利福商业项目
Life Style Shopping Mall, Shenyang

总建筑面积	368,538㎡
设计期间	2007/6~2010/12
建设期间	2008/12
竣工年月	2012/12

本项目的基地位于沈阳市中心的沈河区,南侧距离著名的沈阳故宫230m,并与故宫步行街的中街路连接;北侧邻接市主干道的延长规划道路北顺城路和规划中的50m宽的绿化带;东侧为正阳路;西侧靠淡泊巷。总体规划面积约71408m²,基地性质定性为现代服务中心区。

基地交通便利,东侧有建设规划中的1,3号线地铁站点,今后的地铁车站考虑与地下商业动线连通,形成地上,地下连续的商业,服务,观光,休闲的现代时尚的Shopping Mall.基地内有既存的商业设施:太阳广场(占地面积:11780m²),需进行外立面改造;另外南侧用地中央的荟华楼根据要求要保留(占地面积:2250m²);基地北侧隔路相望的住宅区今后将被拆除并规划为绿化带。由于沈阳故宫是世界文化遗产的保护建筑,所以由南至北,新建建筑物的绝对高度被限制为18m,24m,30m。

1期利福广场:
1期太阳广场是一个在保留原构造基础的同时将建筑物改造为购物中心的项目。地上建筑面积约为75000m²,地下建筑面积约为25000m²,共7层。在外观上主要由两大要素构成。一个是具有象征意义的玻璃空间,另一个是呈L型的石质体块,通过两种材质的对比组成了其极富特色的外观。

此外，承袭整体建筑体块呈L型的特点，在立面的四周加入了以中国风为主题的L型石质体块，使整体表现出一种极为强烈的立体感。

2期久光广场；

由于临近故宫，所以着重考虑传统外观因素，对于如何完美的结合古今这个课题上，在设计过程中下了很大功夫。特别是故宫原本还有暗喻固定的蒙古包的含义，于是鉴于该点在设计出现了现代版蒙古包和现代版的故宫。倾斜的三角形外墙面，使用石材和玻璃构成。此外，考虑到一期与二期的统一，与一期利福广场相同，在石材墙面中融入了富有中国感觉的设计。

3期娱乐设施及五星酒店；

五星酒店所附带的的娱乐设施，由位于中央的立体购物口心平台所构成。设计商业立面时考虑了"建筑物的印象""内容""象征性"。这个项目位于沈阳故宫附近，需要与周围协调。所以"建筑物的印象"为特别重要。如果表面设计成商业性高的立面（比如：广告比较多的立面）会破坏周围整体气氛，因此建筑物外侧与建筑物内侧采用了不同的立面设计。

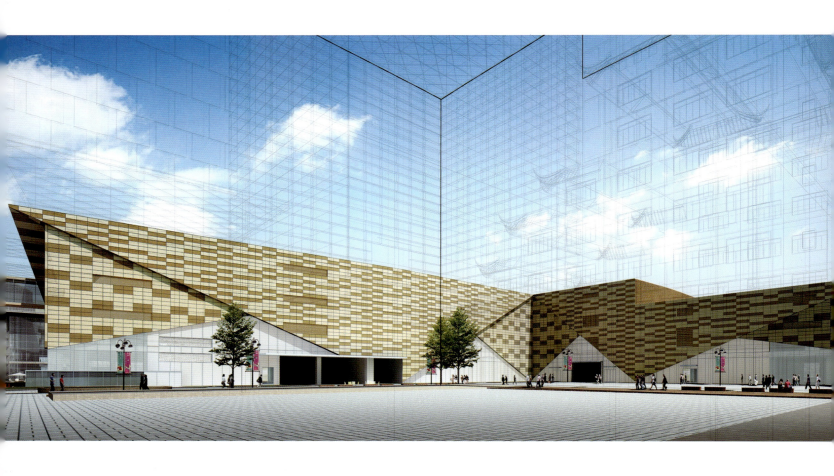

The project base is located in Shenhe District, the center of Shenyang. The famous Shenyang Forbidden City is located 230 meters south of the building site, connecting through Zhongjie Road. Its north is adjacent to the planned extension to the municipal main road along with the 50-meter wide green belt along the City Road; to the north of the site is Zhengyang Road; the west side is adjacent to Danbo Lane. The overall site area is about 71,408 ㎡. The nature of the podium is quantified as modern service center.

The podium has convenient transportation connections; metro stations of line 1 and line 3 are sited in the east part of the development. The subway stations will connect with the underground commercial complex in the future; a modern and trendy shopping mall with business, services, tourism, leisure facilities that continuously connects the above and under ground. The existing Sun Plaza commercial facilities with floor area of 11,780 ㎡ in the podium, requires facade renovation. in addition, the Huihua Building in the center of south plot shall be retained by request (floor area: 2,250 ㎡); the residential area opposite the road in the north of the podium will be demolished and is planned to become a green belt in the future. Shenyang Forbidden City is a protected world cultural heritage building. The absolute height of new buildings in the vicinity is limited to 18 meters, 24 meters and 30 meters from south to north.

Phase I: Lifu Plaza
Phase I Sun Plaza is a shopping center project. The above ground floor area is approximately 75,000 ㎡, and the underground floor area is approximately 25,000 ㎡ with 7 floors in total. The exterior design has two major elements, a symbolic glass room and an L-type StoneBox (stone mass). The distinctive appearance is formed by these two features. As in all commercial developments, the connection with the intersection is the most important area. Here, the semi-exterior space of the "urban gap" and the symbolic glass space are perfectly combined in design, so that the main entrance becomes even more special.

In addition, the overall building mass is L-type. The L shaped stone mass with a China theme is added around the façade with a very strong sense.

Phase II: Jiuguang Plaza
A stronger emphasis on the traditional appearance was attempted since the site is adjacent to the Forbidden City. Much effort was expended in the process of the design for the perfect combination of the ancient and modern. Especially it has been commented that the Forbidden City has the implied meaning of fixed Yurts, the round tents of nomadic peoples. It was decided to design the modern version of yurts and the modern version of Forbidden City. Used as pasted tilting triangles on the walls, the stone and glass seem to create a layered appearance; it is composed of crucial materials inside, it also shows a ray of lightness like floating. In addition, the phase I and phase II are unified, with the design brimming with Chinese cultural references merged into the stone wall similar to Phase I LifePlaza.

Phase III: Entertainment facilities and five-star hotels
The recreational facilities attached to five-star hotels are composed of the multi-story shopping center platform located in the center. The "impression (character)", "contents" and "symbol" were considered in designing the commercial façade. This project is located in the vicinity of the Shenyang Imperial Palace, which requires coordination with its surroundings. Therefore, the "impression (character)" is of particular importance. If the exterior is designed as a highly commercial façade (for example, a facade with lots of advertisements), it will destroy the whole historical atmosphere in the vicinity, so different facade designs are adopted on both outside and inside of the building.

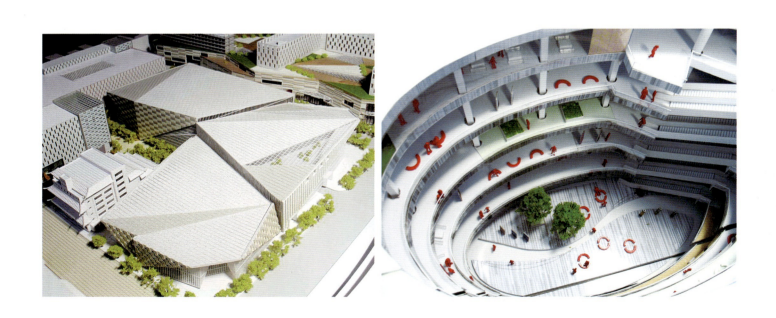

深圳宝安中心区N7综合体
Project N7, Central Business Zone, Bao'an, Shenzhen

总建筑面积	614.451 ㎡
设计期间	2007/6~2007/8(竞赛) 2007/8~2008/10(方案设计)
配合设计	中建国际设计有限公司

深圳市宝安区泰华N7项目取名"城上城"，位于深圳市宝安区新安一路西北面，宝安大道与新湖路之间。用地分为南部公建商业区（以下称作"公建区"）和北部住宅区两个区域进行规划。

公建区包括四座风格各异的建筑：200m的泰华大厦（包括五星级酒店和智能办公楼）、城市创意大厦和两座150m的公寓。四座大楼彼此呼应，围绕着热带山水的低层商业展开。它反映未来的商业发展模式——环境生态型娱乐商业综合体。两条轴线将南北设施连成一体：一条自然轴线，它将庭院珍珠串联在一起；一条体验轴线，将不同风格和内容的设施连接。

它的景观设计也紧紧围绕商业中庭：利用水、绿、光以及人的活动，在建筑物内形成各种各样的中心空间。从住宅栋方向流过来的水流途中，有个由水和绿构成的螺旋形特别空间，使人的活动形成立体连接。一个开阔的可以体验到连接天与地的特别世界。

121

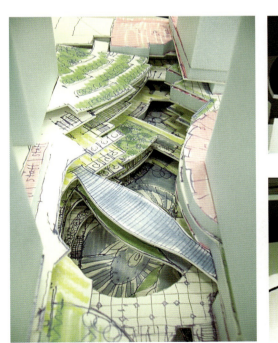

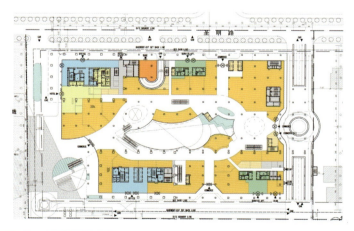

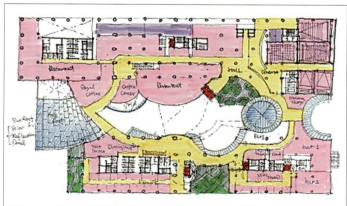

Taihua project N7, Central Business Area, Baoan is called "a City above City". It is located in Baocheng N7 area, Bao'an, Shenzhen City between Xinhu Road and Baoan Road. The site is planned to be a public construction business district in the south (hereinafter referred to as "public construction area") and a residential area in the northern part.

Its public buildings include four different styles of buildings: Taihua building of 200 meters (including a five-star hotel and intelligent office buildings), the urban creative building and the two 150-meter apartments. The four buildings echo each other around the lower business area with tropical landscape. It reflects the future commercial development model – an environmentally and ecologically conscious entertainment and business complex, representing a new chapter for urban development with different appealing aspects from entertainment, environment, society, grade to culture. The gardens at different heights located in the public building area and residential area from south to north are like bright pearls, emitting the special charm of nature and pleasant outdoors. At night, it emits charming lights with a colorful atmosphere. The two axes connect the northern and southern facilities into a whole: a natural axis stringing the yard pearls together; an experience axis connecting the facilities with different styles and contents.

The landscape design is not only the self-discipline of buildings patio exists, but also the internal space outside the existing building. It is an open and special world in which one can experience the connection between heaven and earth and going through the green trees to the water plate. The various central spaces are formed inside each building by using water, green, light and human activities. The water flow from the direction of the residential buildings creates a particular spiral space of water and green, connecting activity in three dimensions. It is open enough to make people feel the special world connecting heaven and earth.

中航昆山城
Zhonghang Kunshan City

总建筑面积	830,000 ㎡（竞赛） 890,000 ㎡（深化）
设计期间	2009/10~2009/11(竞赛) 2009/11(深化)

中航昆山城中体现了我们一贯的环境综合设计手法，通过绿色走廊将完全不同的功能分开，进而提升内部空间结构层次，改善现有城市环境。手法不是单纯的手法，而是依托于用地要素，结合节能技术，搭接不同功能之间的内在联系。

区域规划的主结构围绕昆山商业轴、中航城绿色走廊绿轴和休闲水轴展开。由于用地分为三部分，本次建设强调的是一次规划，分期建设。降低风险，逐步提高用地价值。

· 分期建设思路
一期住宅建设规模为10万㎡。同时建设地下车库。在面向萧林路，建设昆山家居超市2.4万㎡，包括地下一层。

二期打造A7地块，东侧完成昆山三里屯，形成独特商业氛围，同时继续发展北侧住宅区，由于接近河道和深化景观，经济效益要超过一期。

三期打造面向萧林路的昆山中航街，通过街景映射城市发展蓝图。

A6地块建设住宅25万㎡和6000㎡18班幼儿园。精品住宅依托通过造势，造景后的平台，远远超出一期建设住宅的价值，打造中航城的终期蓝图。

· 体现洄游性，连续展开的商业格局
A5地块布置昆山精品百货作为A5标志性建筑。观景电梯沿东西两侧布置，东侧逐步退台，形成台阶状绿堤。

A6是昆山水商城，核心商业设施。采用具有明确洄游动线的商业格局，围绕流水展开，在东西两端布置主力店铺，与西端百货北侧紫竹路底商连接。

A7包括集中商业、超市和昆山三里屯。采用小店铺，通过连廊沟通的布局，利于各自招商。

· 尊重住户利益的住宅规划布局
带状分布是中航城的特色，布局清新，消除围合，将中航城的影响力辐射周边，带动周边，营造健康氛围。严格依照日影计算，经过反复的调整分析，满足高容积率同时，保证大寒日三小时日照的要求，体现规划严谨性。沿河岸没有布置超高层塔楼，而是注重河岸尺度空间，逐步向内侧退让。

中航城不追求过度夸张的表情，而在于给人们体验的空间和时间。这也是我们认为新城再开发时的设计准则之一。

Zhonghang Kunshan City embodies our consistent design approach to environment integration. The completely different functions are separated by the green corridors, enhancing the levels of the internal spatial structure and improving the existing urban environment. This approach is not simple to carry out. It relies on the land combined with energy-saving technologies overlaying the intrinsic link among different functions.

The main structure of regional planning is composed of the Kunshan commercial axis, the green axis and recreational water axis of Zhonghang City Green Corridor. Since the land is divided into three parts, the project places emphasis on a single unified planning stage followed by phased construction to reduce risks and gradually increase the land value.

- Summary of phased construction

The 1st phase: Residential construction of 100,000 square meters together with the underground garage.Kunshan Home Supermarket with 240,000 square meters including the basement.

The 2nd phase: A7 block shall be built;Kunshan Sanlitun shall be completed on the east side to form a unique business atmosphere; Meanwhile,we shall continue to develop the northern residential area. The economic benefits exceed the 1st phase since it is near the river and landscape.

The 3rd phase: Kunshan Zhonghang Street will be built facing Xiaolin Road,a blueprint for urban development shall be reflected through street landscape.

250,000 square meters of residential units and a kindergarten of 18 classes with 6,000 square meters shall be built in block A6. The quality residential area is expected to greatly exceed the value of the first phase residential construction, providing a solid financial base for future development of Zhonghang city.

- The business system will provide for continuous expansion, reflecting the transitional character of the site during development.

Block A5 will have the Kunshan boutique department store as the landmark. The scenic lift is arranged along the east and west sides. The Green Bank on the east side is formed by a stepped landscaped roof terrace.

Block A6 is Kunshan Water Store City and the core commercial facilities. Main shops are arranged in the east and west ends, and department stores are in the west and the final commercials units on Zizhu Road in the north. It is integrated along its whole 300 meter length. Kunshan Arowana is designed in the middle, becoming the symbolic in migration. From the 1st floor underground to the 5th floor above ground, different displays are arranged with crystal scales to bring wealth and good fortune.

A7 includes compact business, supermarket and Kunshan Sanlitun. The use of small shops and the communication spots through corridors is designed to promote investment.

- It also respects the interests of residents

Zonal distribution is a distinctive characteristic of Zhonghang city, with clear layout; the influence of Zhonghang city will permeate throughout the city and promote creation of a healthy atmosphere in its surrounding areas. Exact calculations of shadows cast by sun have been conducted to ensure more than 3 hours sunshine is available on the day of Dahan (a very cold day in Chinese lunar calendar). The meticulous planning has achieved an extremely high volume rate after nearly a thousand adjustments. No high-rise towers are arranged along the river banks. The scale and space of the river bank is respected, gradually give way towards the inside.

Zhonghang Kunshan City refrains from over designed excesses and gives people the space and time for a personnel experience. This is the principle design criteria for the re-development of the new town.

威海宝泉中心（威高广场）
Weihai Baoquan Complex （Wei Gao Plaza）

宝泉广场（原名大操场改造项目）围绕着航路的理念，我们提出带状城市发展格局与主题轴线的理念，塑造环翠楼——刘公岛标志性城市景观轴线，打造连接公园路商业风情街，形成集商业、洗浴、娱乐、高端公寓式酒店、地下空间于一体的威海核心区大操场开发中心。

宝泉广场总体规划和建筑设计围绕锦绣大道展开：

①重视城市整体结构，塑造锦绣大道的城市新轴线。打造环翠楼－刘公岛的锦绣大道，突出"航路"理念。锦绣大道代表了威海城市发展进入了新阶段，也代表了一代代威海人民和建设者为城市发展添砖加瓦，贡献力量。这条锦绣大道连接了自然和历史，具有文化象征意义，和花园路将来打造的步行商业文化街，直接联通宝泉广场以及未来的雕塑公园。

②围绕航路展开的标志性城市商业综合体。建筑围绕五个标志性的广场——盛装广场，欢迎广场、五彩缤纷、海波盎然、威海景湾，依托代表性的自然环境；实现可以对应将来变化和发展的具有灵活性的整体规划；将文化、艺术与娱乐完美地融入到消费中，为市民们带来全新的体验文化。

③营造绿色环境。最大限度地塑造从地下到地面到屋顶的连续的绿色环境，将绿色融入到商业细节之中，让人们感受到山海自然魅力，将天光云影绿意清流融合到建筑空间与景观中，利用种种海滨自然要素陶冶心灵，将环翠楼的绿色和刘公岛的海面连接，承接山风和

总建筑面积	360,000 ㎡
设计期间	2008/11~2011
建设期间	2009/7~
竣工年月	2012/12
景观合作设计单位	有限公司Earthscape
照明合作设计单位	株式会社LIGHTING M
配合设计	上海大境事务所、威海市建筑设计公司
获奖经历	2011年上海建筑学会商业建筑创新奖

海风,融合蓝色和绿色。

④重视地下空间,开放地下商业,强调地下地上一体化商业流线。重视当地气候特征,选用最优化的生态节能技术来创造低能耗、高便利性的绿色环保建筑。利用下沉广场,坡道,中庭等,营造宜人地下商业空间,围绕回路展开的商业轴线,提升地下商业价值。通过巧妙设置地下广场,不仅将自然风引入地下,同时有效提高防灾能力。

Baoquan Complex (formerly known as the playground renovation project) is a romantic composition between mountain and sea. The main concept is sailing routes, and shows colorful pattern facing the Liugong Island. It is proposed to shape the iconic urban landscape axis of the Huancuilou Liugong Island through the complete transformation of the playground lot, under the concept of belt city development patterns and themes axis and also to connect with the business area on Park Road. It is proposed to form the playground development center as the core area of Weihai, integrated with commerce, bathing, entertainment, up-market apartment hotel and underground space.

The overall plan and architectural design of Baoquan Square is around Fairview Park Boulevard:
① Attention is paid to the overall structure of the city, shaping the new axis along Fairview Park Boulevard, developing the beautiful avenue of Cuihuanlou-Liugong Island, and highlighting the concept of "route". Fairview Park Boulevard represents a new phase in the development of Weihai City. It also represents the contribution of Weihai people and builders to the development of the city generation by generation. Fairview Park Boulevard connects nature and history with cultural symbolism, the Baoquan Square and the future sculpture park together with the pedestrian to be built on the Garden Road.

②Baoquan Square is a symbolic urban commercial complex around the sailing route. The building includes five landmark squares –the Well-dressed Square, the Welcomed Square, the Colorful and the Exuberant Seawave, Weihai Bay relying on the representative natural environment; Flexible overall planning which can cope with the changes and future development is envisioned. Cultures, arts and entertainment are perfectly melded with commercial activities to provide the citizens with a grand new cultural experience.

③Create a green environment. Create a continuous green environment to the maximum from underground to aboveground and the roof. The green is melted into business details; let people feel the natural charm of mountain and sea; the sky, clouds, green and water are integrated into the architectural space and landscape; the various natural seaside elements cultivate the mind and comply with the features of Weihai; the green of Huancuilou and the sea surface of Liugong Island is connected; the breeze from mountain and the sea are blended with blue and green.

④ Careful attention to the underground space, open the underground shops, and emphasize the commercial flow lines of integration and underground and above ground. Attention shall also be paid to local climate characteristics. The optimal ecological energy saving technologies shall be adopted to create low-power and convenient green building. The sunken plaza, ramps, patio etc. shall be used to create a pleasant underground commercial space and expand the commercial axis around the loop to improve its underground commercial value. Through ingenious design of the underground plaza, natural wind will be introduced into underground spaces, and also improving disaster prevention capability.

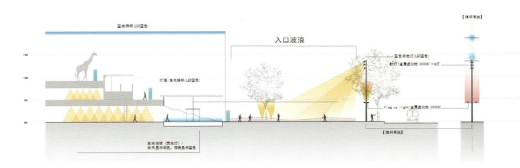

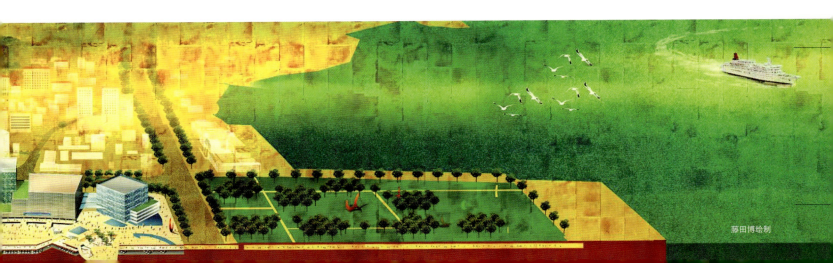

藤田博绘制

日本北九州河岸广场
River Walk Kitakyushu, Japan

总建筑面积	162,473㎡
设计期间	1999/04~2000/03
建设期间	1999/10~2000/03
竣工年月	2000/03
设计监修	FJ城市开发株式会社
合作设计	The Jerde Partnership Inc.
摄影	小林研二摄影事务所
获奖经历	2001年美国建筑奖、北九州岛市都市景观奖、优良消防防灾系统（消防长官表彰）、空调卫生工学会奖（振兴技术振兴奖）、日本最佳停车场奖

北九州河畔广场由5个几何造型的建筑组成，这5个几何造型的建筑代表5个地区，象征着凝为一体的北九州市。从远处眺望，这5栋建筑宛如静态的艺术造型，但是从近处看去，每个几何造型都充满不同的张力与紧张感，可以使人感受到张扬怒放的活力。

除了造型，色彩也是北九州河畔走廊的重要设计元素，给予来访者独一无二的印象与感动。每个建筑外墙的色彩使用及其大胆、充满现代感，同时又融合了日本传统的色彩美学。高层栋的灰黑色、大厅栋的红色，其设计灵感分别来源于传统的日本瓦和日本漆器。中型大厅栋的浅茶色，影院、停车场大楼的金黄色则分别象征着大地的颜色和收获季节麦穗的颜色。低层裙房的白色为日本传统白土墙的颜色，代表生命的纯粹度。

The architectural form of the building represents 5 geometrical shapes that represent each particular region. On the other hand, it was treated as a symbolic of Kita-Kyushu. From afar, it looks like a tranquil objet in repose, but upon approaching the building a dynamic and exciting experience becomes manifest.

Color was also a big theme in the design of the building and leaves a striking and unique impression on visitors. Each building employs a bold and modern color, while keeping in touch with the esthetics of the traditional Japanese color palette. The tall tower is a grey black like the traditional Japanese tiles, the main hall is in a lacquer red, the central hall is a light brown of the natural earth, the cinema and parking building is a yellow that is reminiscent of the rice field's right before harvest, and finally the low-rise block is a plaster white that expresses the theme of purity.

日本桥一丁目大厦
Nihonbashi 1-chome Building (COREDO), Japan

总建筑面积	98,368㎡
设计期间	2000/05~2001/07
建设期间	2001/08~2004/02
竣工年月	2004/02
合作设计	Kohn Pedersen Fox Associate PC
	东急设计咨询株式会社
获奖经历	建筑业协会奖、GOOD DESIGN奖、
	日经优秀产品赏、大理石建筑奖、
	照明普及赏、
	SDA奖（社团法人日本标识设计协会奖）

矗立在立体城市轴交点的高层建筑

毗邻东京日本桥，共20层，高达121m的多功能大厦。坐落在作为自银座延伸出来的商业轴的中央大街与大手町和作为连结兜町的金融轴的永代大街的两个重要城市轴上，西侧与南侧相接的高层大厦林立的街区（A街区）与由它身后的广场和低层餐厅建筑聚集的街区（B街区）构成。

交易楼层为了确保3000㎡的形状规整的办公区，将垂直方向的服务系统集中在北侧，从而在南北狭窄的基地中实现有足够进深的办公空间。高层办公区与石材贴面的服务核心筒以及南侧玻璃覆盖的办公空间形成视觉上的明确分区。

沿永代大街东西方向延伸的南侧弯曲的玻璃面，由于两端刻有深深的沟槽，感觉好似游离于办公体量之外。

在高层办公楼下部嵌入面向日本桥十字路口的玻璃材质的5层体量，内部为商店和大学研究生院，被落地玻璃墙和突出的排列细密的垂直玻璃肋所覆盖。玻璃面相互交错，如同三棱镜般折射出纷飞交织的色彩和光感，形成充满魅惑的外观。玻璃盒子的高度为31m，与周边原有建筑高度相协调。

High-rise Buildings Standing at the Intersection of Urban Axes

This is a 20 story 121m high multi-purpose building adjoining Nihonbashi, Tokyo.
It consists of a high-rise block (Block A) and a low rise restaurant block (Block B) and plaza to the rear, on the west and south sides of two important urban axes: the commercial axis of Chuo Dori extending from Ginza and the financial axis of Eitai Dori that connects Otemachi and Kabutocho.

In order to provide a regular shaped office floorplate of about 3,000m2 for the trading floor, the services system in the vertical direction were concentrated on the north side, to create an office space with sufficient depth even though the site was comparatively narrow in the north- south direction. The high-rise office part was clearly divided visually into a service core part covered with stone, and an office space covered with glass on the south side.

The curved glass surface on the south side extending east-west along Eitai Dori appears to be separated from the office volume by deep grooves at both ends.

A 5 story glass volume is inserted into the bottom part of the high-rise offices facing Nihonbashi Crossing. Inside are retail shops and a graduate school, fully covered by a glass wall with closely spaced vertical glass fins. This mixture of glass surfaces creates an attractive prism-like colorful feeling. This glass box is 31m high from the top, the same height as the surrounding existing buildings, to harmonize with the urban landscape.

142　越南外交部大楼
Headquarters of the Ministry of Foreign Affairs, Vietnam

144　越南河内政党委员会总部
Hanoi Party Committee Headquarters, Vietnam

148　越南财务部数据中心
Financial Data Center Competition, Vietnam

150　日本岩手县县民情报交流中心
Iwate Citizen Information Exchange Center, Japan

Government Buildings
政府行政建筑

越南外交部大楼
Headquarters of The Ministry of Foreign Affairs, Vietnam

新的外交部大楼设计的基本概念,是用于国际间交流的新设施。至于功能职责,其有责任与具有不同背景的不同地区、文化和个人建立联系。交流的主题很自然地适合于本项目。鉴于此,交流的概念就不仅仅是一种静态标志,更是一种具有催化作用的哲学原理,外交部大楼已经成为了越南社会主义共和国的海外形象所在。同样,项目设计美观、宁静,周围被大量的自然绿色景观包围。

越南外交部的3个概念:

1、网络:为了能够实现最大的场地利用率和创建最大的交流空间,建筑物的布局注重周围环境和建造计划的关系。通过采用最新式的输送系统和外交档案的放置情况以及中央区域内的控制中心,可以通过一个自动化的系统来控制信息。

2、安全:外交部大楼作为交流场所就必须具有周全的安全系统。在发生紧急状况的情况下,设施必须提供3个级别的安全性:人员安全,信息安全,基础设施安全。

3、景观:景观设计与交流空间的功能和范围有关,旨在为交流活动提供最佳环境,它同样也是人与自然之间的一种比较亲近的交流关系。景观设计主要包括3种类型的景观区域:外部设计、内部设计和中间区域。外部设计包括建筑物外观风格的一致性,而这种一致性可以把建筑物与城市建筑连接起来,从外面看,建筑物就像地处自然景观中一样。内部设计主要包括建筑物中心的中央庭院,是一种面积比较大的绿色植物波浪起伏的景观带。中央庭院通过走廊把整座建筑连接成一个整体。同样,通过外围设计和连续的景观设计,场地上波动起伏的绿色带也创造出一系列意想不到的空间。建筑物的中间区域包括高楼之间的实际连接通道,和供人员会面、休憩及交流的露天平台和花园。

总建筑面积	198,682㎡
设计期间	2008/3~2008/5

As a modern nation on the international political stage, Vietnam and the Ministry of Foreign Affairs is an emerging identity striving to create valuable and varied relationship with the inner and outer world. As the prime concept for the design of the new MOFA building, it is appropriate that the new facility becomes a platform for universal exchange. In its functional role, it has the duty to create links with various regional, cultural and individual bodies from different backgrounds. Therefore, Exchange naturally becomes the main theme for this project. Whether it is the exchange between people, political parties, ministries, governments or nations, the design of the new facility excels in becoming the ultimate exchange headquarter. With this in mind, the notion of Exchange is not only a static symbol of an activity but rather a catalytic philosophy to propel Vietnam as a leading nation internationally in the near and upcoming future. In addition, MOFA being a national body representing the Vietnamese Government worldwide, it becomes the image of the Vietnam Republic overseas. As such, the design of the facility shall be beautiful, serene and abundantly surrounded by natural green landscapes.

MOFA's 3 Concepts:
1. Network: The layout of the buildings is conscientious of the surrounding environment and the building program, in order to maximize efficiency of the site usage and the creation of exchange spaces. With the use of the newest conveyance system, and the placement of the diplomatic archives and control center in the central area, information can be controlled via an automated system.

2. Security: An indispensable requirement for an adequately designed MOFA Building as a place for exchange is a well thought out security system. In the instances of emergencies, the facility must provide proper security at 3 levels: 1) People, 2) Information, and 3) Infrastructure.

3. Landscape: The landscape design is relational to the function and scale of the exchange space to provide the most appropriated dynamism and enlivenment for exchange activities, and also a close exchange relationship between people and nature. The landscape design is mainly composed of the following 3 types of landscape zones: Outer Design, Inner Design, and In-Between zones. The Outer Design consists of a continuity of the building façade's pattern therefore bringing connectivity to the buildings with the urban fabric. Viewed from the outside, the buildings seem as though they were spawn from the landscapes. The Inner Design consists mainly of a central courtyard at the center of the premises, designed as a vast undulating green landscape. This central courtyard provides the whole site with a whole entity by visual connectivity. Also, through the design of an enveloping and continuous landscape design, undulation on the site creates a series of unexpected spaces. The In-Between spaces between the buildings consist of physical linkage between the towers, yet also serve as an exchange space between guests and office staff. In consists of a deck and garden where users can meet, relax and communicate.

越南河内政党委员会总部
Hanoi Party Committee Headquarters, Vietnam

总建筑面积	19,960㎡
设计期间	2007/9~2007/11
获奖经历	国际设计竞赛一等奖

该 项目是2007年举办的国际设计竞赛。日本设计在竞赛中获第一名。设计理念包括建造一个新型现代化的河内市地标，以庆祝其成立1000周年。项目位于河内中心区域中的一块梯形场地上，西邻Ngo Quyen大街，南邻Le Lai大街，东邻Tong Dan大街。区域中主要包括政府办公室和金融机构。

The design concept consists of creating a new modern landmark for Hanoi city celebrating 1000 years of its foundation. The project lies in a trapezoidal shaped site located in central Hanoi bordered by Ngo Quyen St. on the west, Le Lai St. on the south and Tong Dan St. on the east. The area consists mainly of government offices and financial institutions. Important features of the surrounding area include the major crossing of Le Lai and Ngo Quyen streets.

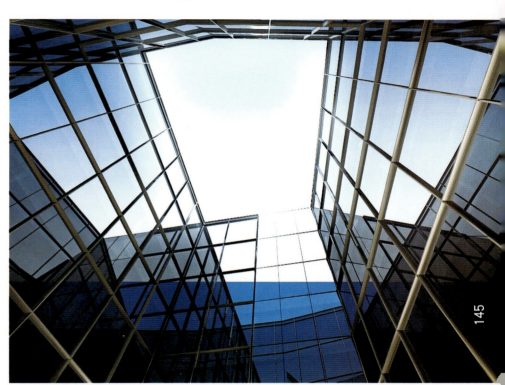

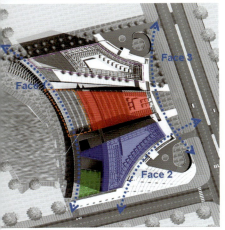

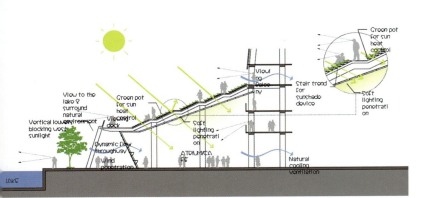

越南财务部数据中心
Financial Data Center Competition, Vietnam

总建筑面积	47,000㎡（一期） 30,000㎡（二期）
设计期间	2010/2~2010/12
获奖经历	国际竞赛第一名

越南河内市财务部数据中心。项目是由越南财政部为越南河内著名的乐高科技园设备基地研究中心的场所主办的一项设计参赛项目。分两期：一期包括含财政部窗口功能的36,000㎡办公楼与约10,000㎡的数据中心以及1,000㎡的能源中心，2期主要为30,000㎡左右的办公楼。

这一综合设施的设计突出了作为建筑群的整体感。3个具有不同意义的象征性的曲线构成了整体体量和形状。

第一个曲线位于基地内部环境最好的湖边，有意识地强调建筑与环境的联系。结合这一曲线布置圆形剧场（阶梯型观众席）和景观通道，作为集会、休息等场所。另外，台阶上以绿化进行点缀，实现与周边环境的和谐统一，同时提高作为休息场所的空间品质。

第二个曲线主要强调建筑的正面性。多数越南人习惯于将开放的主入口正对道路交叉口。因此，在本项目

设计中以道路交叉点为中心形成圆弧形的建筑体量，强调了主入口的正面性。

第三个曲线强调了主干道与本项目建筑群之间的关系。连接主干道的入口部分构成了优美的曲线。巨大的弧形和主干道之间产生的"气穴"空间，从视觉上将人和车辆引导至建筑群。

这些曲线同时将建筑群作为一个整体展现在人们的眼前。

覆盖建筑的玻璃幕墙采用彩釉玻璃，将计算机数据处理中的基本符号0和1随机排列构成数据中心的标志性图形，实现了业主所期望的现代、先进、象征性的建筑。

The project is the entry for a design competition sponsored by the Ministry of Finance for facilities of the research facilities base center in the well-known Legao Hi-Tech Park, Hanoi, Vietnam. The project is divided into two phases: phase one includes 36,000 ㎡ office building, about 10,000 ㎡ data center and 1,000 ㎡ energy center with the window function of the Ministry of Finance; phase two is composed mainly of 30,000 ㎡ office buildings.

The design of this complex highlights the associative perception as a building complex. The three symbolic curves with different meanings carve out the overall mass and shapes.

The first curve is located near the lake enjoying the best internal environment of the complex. It places conscious emphasis on the linkage between architecture and environment. Combined with the curve, the amphitheater (stepped auditorium) and landscaped path are arranged as places for gathering and resting etc. In addition, greenery is planted on the stepped roof to achieve harmony with the surrounding environment. The landscaping also improves the spatial quality as a resting place.

The second curve mainly emphasizes the frontal orientation inherent to the site. The arc-shaped building mass centered on the road crossing was formed, since most Vietnamese are used to having main entrance face the road intersection, to emphasize the frontality of the main entrance.

The third curve mainly emphasizes the relationship between the main road and project buildings. The entrance connecting to the main road constitutes a beautiful curve. The "cavitation" space is produced between the huge arc and the main road. It visually leads people and vehicles into the buildings.

These curves create the lasting image of the building group as a whole.

The glass curtain wall which covers the building use colored glazing; the basic symbols 0 and 1 in computer data processing constitute a graphic symbol of the data center by random arrangement. It epitomizes the modern, advanced and symbolic building which the owners expect to be achieved.

日本岩手县县民情报交流中心
Iwate Citizen Information Exchange Center, Japan

摄影：Ono Studio Ltd.

总建筑面积	45,874㎡
设计期间	2000/09~2002/05
建设期间	2003/03~2005/09
竣工年月	2005/03
其他设计单位	曾根幸一·环境设计研究所 久慈设计
摄影	外观 / Ono Studio Ltd.
	中庭 / 木田胜久/FOTOTECA

这是以促进岩手县民间信息交流为目的的综合性设施，集中了13个功能各异的公共设施。在设计过程中根据每个设施的性格特征，对空间结构进行反复推敲及调整，整个设施被划分为"乐"、"知"、"学"三个空间，实现了富有特色的功能分区。"乐"空间指的是中间层，以信息、展示等为主体，并配有供来访者休息、停留的中庭。"知"空间指的是低层裙房，设有需要大空间的图书馆等行政服务设施。"学"空间指的是高层部分，设有多功能大厅、会议室、县民活动设施等。在高层"学"空间内设有集会设施、功能厅和会议室，采用素色的木材及玻璃，设计成为漂浮在中庭的盒式独立空间，营造出独特的空间特色。

从城市设计的角度来看，aiina突出了水平元素，与强调垂直元素的邻栋建筑"MALIOS"大厦形成鲜明对比。为了强调中庭的开放感，实现图书馆宽敞的书架模数结构，建筑的V字形柱子在高层和中间层采用不同的柱跨 通过在中间层的转换，从结构上解决了柱跨问题。形态富有特色的柱子为连绵伸展的外观赋予韵律感的同时，提升了高层部分的漂浮感。高层部分的玻璃外墙，面向盛岗站西口剧场，打造成迎接各方来客的圆弧造型，同时还可以眺望广为岩手县民喜爱的岩手山，成为具有象征意义的门户型建筑。

摄影：木田胜久/FOTOTECA

Aiina的玻璃外墙采用隔热型Low-E中空玻璃，实现了节能，并且利用挑空中庭的拔风效果，实现了自然通风，形成呼吸式外墙，同时利用地下基坑的地热箱，减少新风负荷，降低了夏季及冬季的空调运行成本。

在"学"空间设有可长时间利用、面积相对小一些的房间，这些房间容易受到窗周部的自然光和温度的影响，因此双层幕墙采用平板玻璃和隔热型Low-E玻璃，利用在双层幕墙之间形成的拔风效果进行自然通风，并在双层幕墙上设置大型纵向百叶，调节日照，降低室外负荷。双层幕墙的最高高度虽然有5m，但是通过采用铝合金压型材，实现了轻量化，可进行手动调节。同时通过设置肋板，确保玻璃幕墙的平整度，使自然光在室内营造出柔和的光线。采光顶采用装有太阳能发电板的建材，向来访者展示着积极的环保态度。

This building is a complex that combines 13 public facilities for the information exchange of citizens. The design of the facility put much thought into the positioning, space and organization of the functions. In the mid-rise portion is the central atrium which acts as the main gathering space for information and exhibitions. In the low-rise portion the library, administration services are located and in the high-rise portion the multi-function hall, meeting room and citizen's activity hall are located, zoned according to the requirements of each function. Also the various gathering halls are located in the high rise tower. The large meeting hall acts as the main feature floating in the middle of the large atrium clad in natural woods and glass.

The planning for the center contrasts with the design of the adjacent MALIOS center by emphasizing horizontal features. V shaped columns were designed to achieve openness in the atrium and spaciousness for the bookshelf modules of the library. To do so, the column spans change at the mid-levels. The uniquely shaped columns are placed across the facade giving the appearance of a continuous rhythm, accentuating the lightness of the high-rise portion. The glass wall of the high-rise portion facing the west plaza of the station, open up as an arc-shape welcoming the visitors, as well as present a symbolic façade facing Mount Iwate.

The exterior wall is built with a low-E insulated layered glass and the atrium allows for natural ventilation through a natural draft, breathing exterior membranes for energy efficient design. Furthermore, use of an underground pit and a system of cool-heat trench technology reduces the energy loads on air conditioning in both summer and winter, lowering running costs.

On the east-west side, floated plate glass and insulating Low-E glass are used to make a double-skin structure. The east-west sides have heavily used small rooms that are vulnerable to the strong light and temperature from the perimeter zone. The inside of the double-skin structure is an open void that allows for a draft to flow in and bring in natural ventilation. Also, large vertical louvers were put in place to control the sun light and shade of these orientations. To reduce the structural load of the exterior wall, the louvers were designed in aluminum, reaching 5m in maximum height, and are manually operational and moveable. The louvers also create a pattern of light and shade on the facade. Finally, the material used along the skylight is covered with solar panels which are visible to the passersby, emphasizing the eco-technology that is implemented throughout the building.

154 深圳龙岗中心医院
Longgang Central Hospital, Shenzhen

160 台州恩泽医疗中心
Taizhou Hospital

164 上海浦东东方医院
Pudong Eastern Hospital, Shanghai

168 日本金泽医科大学病院
Kanazawa Medical University Hospital, Japan

170 日本岐阜县综合医疗中心
Gifu Prefectural General Medical Center, Japan

Medical Facilities
医疗建筑

深圳龙岗中心医院
Longgang Central Hospital, Shenzhen

总建筑面积	55,000㎡(地上)
设计期间	2005/3~2005/6(竞赛) 2005/7~2006/12(设计)
建设期间	2005/12~2008/12
竣工年月	2008/12
配合设计	深圳新城市建筑设计有限公司
摄影	深圳市匠力摄影设计有限公司

深圳龙岗区中心区的龙岗中心医院（深圳市第九人民医院）是2004年经过国际投标中标的项目，2008年12月24日正式投入使用。在投标中方案由于造型独特，手法简洁，医疗建筑流线明确等博得了好评。特别是设计中采用的自然通风设计手法不仅体现了绿色医院建筑设计，也引起了广泛的关注。

龙岗中心医院采用的是接近正方形的平面，很好地处理了西侧面向主要道路的城市关系。建筑垂直功能分布则采用集中式布局：门诊位于低层，医疗技术位于中层，病房位于高层。口字形病房楼层布局的特点在于，标准病房层的病房围绕着中庭布局，核心筒分布在中庭周边。为了证明架空层和中庭的实际通风效果，日本设计的设备工程师进行了计算机模拟实验，验证通风和上升气流所产生的实际效果。通过实验证明，设计中采用的中庭和病房中采用的生态通风井都可以产生足够的上升气流，达到节能和改善病房层内部环境的作用。整个实验的过程都是通过计算机和专用软件进行的，随着建筑数据的修正和外围条件数据的细化，需要反复地继续测算。通过实验结果可以清楚看到除了中央占据主要位置的通风中庭外，主要病房靠近走廊一侧结合设备管道都布置不同的生态通风井。这些生态通风井为病房提供了循环气流，解决了高层建筑室内外气压

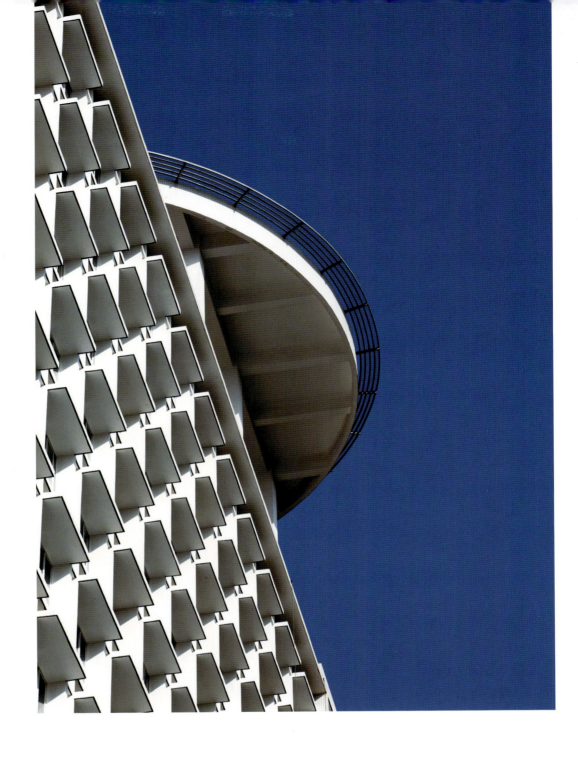

问题。它和病房通风区域、非病房通风区域都是通过架空的六层直接导入气流，同时利用与屋顶升高的排气口之间高差以及温度差，产生上升气流。

中庭和病房内的自然通风系统相互独立，形成了相对独立的病房区和公用医务区两个空气循环区域。公共医务区包括走廊、护士站等公共区域，通过对外通风的活动室和走廊部分，形成直接的进风口，从病房层的中央通风采光井直接排出。病房区通风系统包括病房区域从充分向外展开的病房窗户和外部阳台形成充足的进风通道，宽度达到3.3m，从最上层的生态排风口排出。基本上可以解决公共区和病房区的通风换气的需求。

龙岗中心医院主要病房面向南侧和东侧，西侧作为公共空间，包括电梯厅和卫生间，这样可以避免正方形平面造成的病房布置不利的问题。为了避免西晒，建筑的西立面采用了外挂混凝土板，既有效地改善了西晒对于内部病房和门诊的影响，同时形成了虚实结合的主立面设计。在外挂混凝土的后面是两组避难疏散楼梯和部分室外空调机的放置架，它也起到了一定的遮挡作用。同时医院主入口也位于医院主楼的西侧，进入主入口直接来到阳光大厅。大厅位于高层的南侧，从西侧向东侧贯通，向中央庭院展开。各个主入口上面设计了大雨棚，突出了城市之门的标志性视觉效果。

The Longgang Central Hospital in the central area of Shenzhen Longgang District (the Ninth People Hospital of Shenzhen City) is an international tender project in 2004. On December 24, 2008, it was formally put into use. In bidding, because it is unique, simple, clear flow lines of medical buildings, it won so much praise. Particularly the natural ventilation design methods used in design reveal the green hospital building design, but also attract wide attention.

Longgang Central Hospital adopts the nearly square plane; it is very impressive in Longgang District. It deals with the west side facing the main roads. Vertical function distribution of the buildings adopts centralized layout: section for outpatients are in lower part, medical technology in the middle, ward at the top. "Kou"- shaped (the Chinese character) floor layout of the ward is characterized by: the wards on the standard ward floor are around the layout of patio, the core tube is located in the surrounding of patio. To prove the actual ventilation effect of the empty floor and patio, the equipment engineers of Japan design conduct computer simulation experiment to test the actual results of ventilation and increased air flow; it is proved by experiments that, the patio used in design and the eco-ventilation used in wards can generate enough upwards airflow, to save energy and improve the inner environment on the ward floor. The whole process of the experiment is carried out through a computer and special software. With the correction of construction data and the refinement of the external conditions data, we need to calculate repeatedly. it can clearly be seen from the results that except the center occupies the ventilation patio in dominate position, the main wards near the corridor with equipment pipes are arranged with different ecological ventilation shafts. These eco-ventilation shafts provide wards with air circulation, and solve the problem of indoor and outdoor air pressure in high-rise buildings. Air is introduced in the ventilation areas in wards and the ventilation areas in non-wards areas by the 6th empty floor; meanwhile, regional ventilation through the ventilation area overhead air directly into the 6 floors, while the upwards airflow is produced by the temperature difference and height difference between the

increased roof vents.

The natural ventilation system in patio and wards are independent to each other, forming two air circulation areas: a relatively independent wards area and the public medical area. Public medical areas include corridors, nurses' stations and other public areas; the direct ventilated activities rooms and corridors form the direct air inlet; it is discharged from the central ventilation well in the center of ward floor. The ventilation system in the wards area includes the sufficient ventilation channel from ward windows and external balcony in the ward area with width of 3.3 meters, discharged from the top ecological discharge port. Basically, the ventilation requirements can be solved in public areas and wards areas.

The main wards in Longgang Central Hospital face the south and east, the west is as a public space including elevator hall and bathroom, thus it can avoid negative arrangement of wards caused by square planar. In order to avoid sunshine in the west, the west sides of building adopt plugged concrete panels, which effectively improve the effect of sunshine in the west on the internal wards and outpatient; meanwhile the main facade design is done with the combination of virtual and reality. Behind the plugged concrete panels, there are two sets of evacuation stairs and holders for some outdoor air conditioners; they also play the role as blocking; at the same time, the main entrance of hospitals is also located in the west of the main hospital building; after entering the main entrance hall, one can come into the sunlight hall directly. The hall is at the south of high-rise, extending from west to east towards the central courtyard. A large canopy is designed on each main entrance, highlighting the symbolic visual effects of city gates.

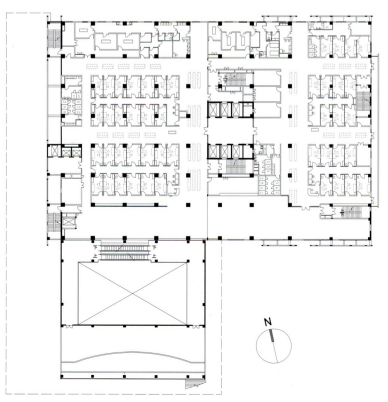

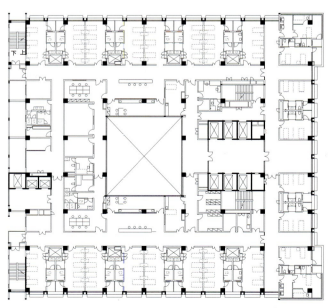

9层：颅脑神经外科、ICU
10层：心胸外科、泌尿外科住院病区
11层：普通外科住院病区
12层：整形住院病区
13层：五官科系统住院病区
九到十三层各层建筑面积都为2502.53m²

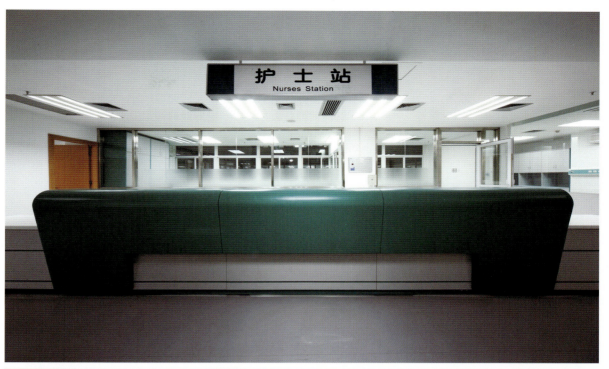

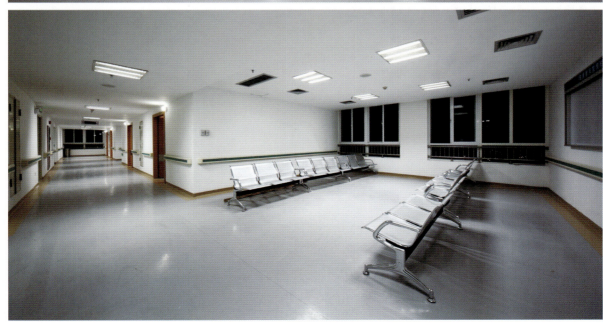

台州恩泽医疗中心
Taizhou Hospital

总建筑面积	131,950㎡（1期）
设计期间	2006/10~2006/12(竞赛) 2006/12-(设计)
建设期间	2007/12
竣工年月	2011/12
配合设计	浙江省建筑设计研究院
摄影	姚文瑜

台州——人杰地灵的海滨城市，属于浙江省城市化格局中的大城市和一级经济亚区中心，是长三角区域16个城市之一。台州恩泽医疗中心历史源远流长，从百年前的恩泽药局，到今天台州惟一的三甲医院。2010年底成立台州恩泽医疗中心集团，将台州医院建设推向新的篇章。

台州恩泽医疗中心将会屹立在城市的交叉点上面，主入口面向北侧，建筑的体型要相应标志性的地位。医院主入口面向北侧国道，建筑面向南侧的恩泽生态绿色广场展开，在强调医院开放感的同时，突出了绿色医院和可呼吸医院的全新理念。

整体总图体现了中国传统文化中院落围合的空间概念，将绿地作为整体的规划中心，体现了恩泽医疗中心以人为本的理念。主要的塔楼分一期、二期、三期建设，呈45°展开，呈现了恩泽医疗中心为社会提供医疗服务的核心地位。周边的道路构成具有强烈凝聚力的格架结构，迅速地到达各个建筑，同时保证了内部步行者空间的独立和舒适性。

恩泽医疗中心的景观规划更多提供给患者自然的空间和可以恢复的环境。建筑的立面构成在体现了中国传统的古典空间构成原则，低层突出连续展开的水平线条，塔楼强调垂直性同时，强调塔楼之间的彼此呼应。

医院服务区包括了医院各种面向社会的配套设施，面向城市主干道，起到了积极的补充功能。主体区包括三栋住院大楼和门诊中央医技等功能，通过围绕生态绿色广场展开的绿色回廊，将不同功能的单体建筑结合在一起。在医院西南角相对独立的设置了传染病区。VIP区在东侧临水地区，设置了包括体检中心，VIP住院大楼，专家公寓在内的VIP区，良好的视野和与自然水体的结合。生活区在用地南侧，集中布置了学生宿舍等居住设施。台州恩泽医疗中心主要患者动线都集中在北侧宏伟展开的大雨棚之下，令患者一目了然。为了解决大量人员就诊，方便患者，设置了专用的区间巴士往返于城市主干道和医院大门之间。

整体规划、分期建设
由于用地面积大，本次建设强调的是一次规划，分期建设。首先一期完成医院建筑主体和整体路网，建设成具有800床规模的大型医院。二期，建设新的VIP服务大楼，对于原有各个病房区进行整合，增加病房400床，达到1200床规模。三期，根据市场情况，建设三期恩泽医院大楼，以及对医技部分的扩建和医院配套设施的完善，达到1600床，甚至更高的大型现代化医疗中心。

医院整体建设具有可调整性，发展性，围绕环境为主题，成为名副其实的可以呼吸的绿色医院。

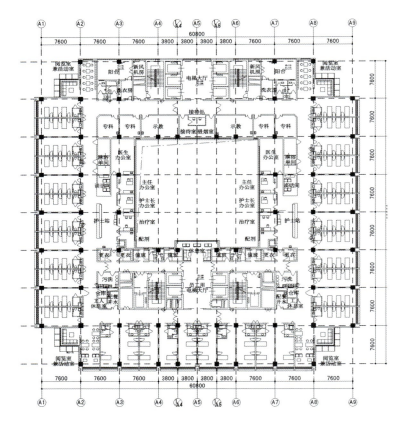

Taizhou – a coastal city full of elite people and resourceful land, a large city in the cultural pattern in Zhejiang Province, the 1st economic sub-district center, one of the 16 cities in the Yangtze River Delta region. Taizhou Hospital is with long history, from the Enze Pharmacy a century ago to the only 3A hospital in Taizhou today. Taizhou Medical Center Group is established at the end of 2010, which promotes Taizhou Hospital to a new chapter.

Taizhou Hospital will be standing in the intersection of the city, the main entrance faces the north, the construction body needs the corresponding iconic status. The main entrance of hospital faces the national road in the north; construction extends towards the southern Enze ecological green square. When stressing the opening feeling of hospitals, we also highlight the brand new concept of green hospital and the hospital which can breathe at the same time.

The overall figure reflects the enclosed courtyard concept in traditional Chinese culture, takes vegetation as planning center of the whole planning. This reflects the concept of people-centered of Taizhou Hospital. The main towers are to be built by phase I, II, III, extended at 45 degrees to cater to the all angles; this shows that Taizhou Hospital's core position which provide medical services for the society. The surrounding roads constitute a frame structure with strong cohesive power; thus we can rapidly arrive at each building, meanwhile we can ensure the independence and comfort of the internal pedestrian space.

The landscape planning of Taizhou Hospital provides patients with more natural space and environment which can be restored. The facade constitution of the building reflects the constitution principles of Chinese traditional classical space. The lower floors make the lines of continuous expansion outstanding; the towers emphasis on verticality and echo between each other.

Hospital service area includes all the supporting facilities of hospital for the society, facing the urban trunk roads; this plays a positive role in complement.Subject areas include three residential buildings and outpatient medical technology center. The green ecological corridor extended around the ecological green square combines the individual buildings together with different functions. Relatively independent infectious disease areas are set at the southwest corner of hospital.VIP area is in the east near water, including the medical center, VIP building and experts' apartment. Good vision combines with natural water. Living area is in the south of land, where student dormitory and other living facilities are arranged. Most patients in Taizhou Hospital are under the great canopy extended in the north. Patients are very clear at the first glance. In order to solve the problems of large number of patients and be convenient to patients, we set up special shuttle bus to and from the main urban road and each gate of the hospital.

Overall planning, phased construction
The land area is great. This building emphasizes one-time planning and stage construction. Firstly, phase I will complete the main hospital buildings and the overall road network, a large-scale hospital with 800 beds. Phase II will build a new VIP service building, integrate the original ward areas, increase 400 beds up to 1200 beds. according to market conditions, Phase III will build the Taizhou Hospital building, as well as expand the medical technology department and improve hospital's supporting facilities to 1600 beds, and even become a more large modern medical center.

The overall construction of the hospital is of scalability, development, focusing on the theme of environment. This will become a green hospital which can breathe in reality.

上海浦东东方医院
Pudong Eastern Hospital, Shanghai

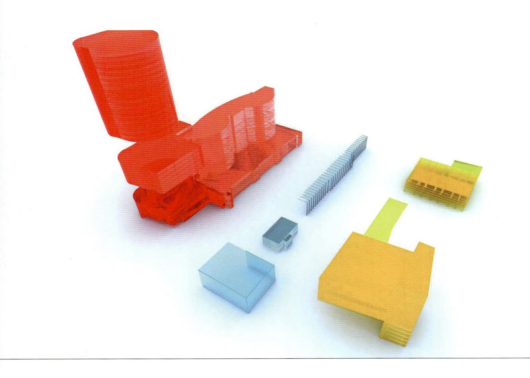

总建筑面积	136,867㎡
设计期间	2008/1~2008/3(竞赛)

浦　东东方医院位于陆家嘴金融开发区中心位置，作为老医院的扩建，新的东方医院将成为一所集临床、教学、科研、预防、保健、康复为一体的现代化国际综合性医院，同时承担区域内健康保健、体检、医疗保健、疾病治疗的社会需求。

新医院建设面向未来，不仅实现高端医院的建设，而且与原有医院整合，实现资源效益最大化，注重经济性、节能性，高效率地完成增建工程。医院相关功能在北侧，包括国际医院、原有医院、国际急救等；南侧是学术交流、办公管理和健康检查。体现出北侧为医疗区（患者），南侧为服务区（非患者）的划分。同样，西侧是新区：国际医院、学术办公中心；东侧是东方医院和扩建的停车场。新的医院与原有医院形成L形布局，通过生态长廊连接不同空间，将不同功能巧妙地联系在一起。

医院服务区
包括了医院各种功能，面向城市主要干道，与将来建设的地下铁和公交车站

相邻，便于提供社会医疗服务。包括国际医院、国际急救、重大保障和原有东方医院的所有医疗功能。

保健办公区
集中了不同需求的保健功能，干部保健、高端保健以及办公、医院管理公司和学术交流研究。同时保留向南侧继续扩建的余地。

生态轴线——通过生态长廊将不同功能连接在一起
位于原有东方医院南侧扩建的生态长廊整合了老医院的门诊入口、急诊入口和小儿科入口，同时连接了西侧的新医院和南侧的停车楼，保证患者能够拥有舒畅的就诊环境。它实现了从东到西的联通，以及北侧园区和南侧绿地的贯通。其本身具有各种节能生态技术，由于绿地融合，形成了名副其实的生态轴线。作为整体布局中心的绿色庭院是院内重要的绿地空间，与绿色交流大厅形成连续的绿色轴线，构成医院的核心空间。

由于东方医院是改扩建项目，它的整体规划和分期实施至关重要。我们通过现场调查和研究，提出了缜密的分期建设和扩建计划：新建的东方医院占据医院西侧，与原有医院形成L形布局，在东侧建立自走式停车库。通过生态走廊，连接不同功能，同时围合成朝南展开的中央绿地，它与能源中心成为新建医院的布局核心。医院功能布置在北侧，南侧是管理和学术、健康检查。考虑到将来东方医院的发展前景，本次扩建必须具有先见性，避免原先能源中心扩建时造成的不必要的问题。

这个项目是日本设计参加的国际竞赛，虽然在竞赛评审中取得了专家的一致好评，但是由于特殊客观原因没能最终实现。

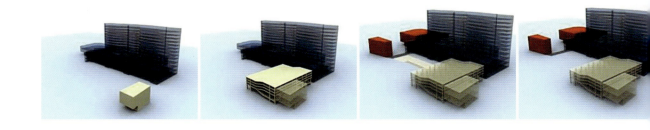

Pudong Eastern Hospital, Shanghai East Hospital in Pudong Lujiazui Financial Area in the center, as the old hospital's expansion, the new East Hospital will become a set of clinical, teaching, research, prevention, care and rehabilitation as one of the modern international general hospital, and to assume the region health care, medical, health care, disease treatment needs of the community.

The building of new hospital is facing for the future; we not only achieve high-end hospital construction, but also integrate with the existing hospital to maximize the efficiency of resources, focus on economy and energy saving, and complete newly built projects with high efficiency. The hospital-related functions are in the north, including the International Hospital, the original hospital, international first aid, etc.; in the south, there are academic exchanges, office management and health checks. This reflects the planning of the medical area (patients) in the north, service area (non-patients) in the south. Similarly, the west is a new area: International Hospital, Academic Office Center; the east is eastern hospital and the expanded hospital parking lot. The new hospital and the original hospital are laid out as f L-shape; the different spaces are connected through the ecological corridor; different functions will be linked together in a clever way.

Hospital Service Area
It includes various functions of the hospital; it faces the city's main roads, adjacent to the metro and bus stop in the future; it is easy to provide social care services. It includes international hospitals, international first aid, major security, international hospitals and all the medical functions of the existing east hospitals.

Health office area
It is integrated with health functions for different needs, such as cadres' health care, hospital management companies and academic exchanges. Meanwhile, it shall maintain continuing to expand to the south.

Eco-axis -- different functions are connected by ecological corridor
The expanded ecological corridor, which is located in the south of the original East Hospital, integrates the outpatient entrance, emergency entrance and pediatric entrance of the old hospital; meanwhile it connects the new hospital in the west and parking building in the south, so that patients can be treated in a comfortable environment. It achieves the connection from east to west, northern park and southern vegetation. It owns a variety of energy-saving ecological technologies; the integration of vegetation space forms a veritable eco-axis. The green garden, as the center of the overall layout, is an important vegetation space; the green communication lobby together with it forms a continuous green axis, constituting the core space of the hospital.

As the Pudong Eastern Hospital, Shanghai is a renovation and expansion project; its overall planning and phased implementation are essential. Through field investigation and research, we propose the careful stage construction and expansion plans: the newly built Pudong Eastern Hospital occupies the west of the hospital, forming L layout with the original hospital. We will establish a self-propelled parking garage in the east. The ecological corridor can connect different functions; meanwhile it encloses a central vegetation expanded towards the south; the Energy Center with it will become the layout core of the newly built hospital. Hospital functions are arranged in the north; management, academy and health checks are in the south. Taking into account the future prospects of the Pudong Eastern Hospital, this expansion must be of prevision to avoid the unnecessary problems when the original energy center expands.

This project is designed to participate in the international competitions by Japan Design. Although unanimous experts praise is gained in the competition assessment, it could not be realized eventually due to special objective reasons.

 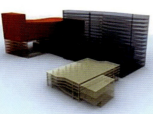 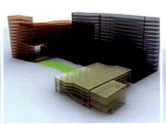  

摄影：SS北陆株式会社

日本金泽医科大学病院
Kanazawa Medical University Hospital, Japan

大学附属医院中央医疗大楼

在医院二次开发中集约中央医疗功能的重建计划。本中央医疗大楼发挥连接先期竣工的西病房楼和原有门诊大楼的作用，并充分重视各设施之间的联系。因此，与西病房楼联系较多的部门被设在同一楼层，由于层高相同，两楼之间的交通联系十分便捷。希望利用裙房与原有门诊大楼相连，这样也便于将来门诊大楼的重建，另外还建设了一座设有坡道和自动扶梯的联络大楼。

Centralization and Coordination of Medical Functions University Hospital Re-development Project Team
This is a hospital re-development project in which the central medical diagnostic and treatment functions were centralized. The central diagnosis and treatment building has the role of connecting the already completed hospital building on the west and the outpatients building on the east, and importance was placed on linking each facility. Therefore departments that were strongly associated with the west hospital building were placed on the same floor, so that efficient movement between the two buildings was provided by having them on the same level. A connecting building having a slope and escalators was provided next to the existing outpatients building in order to provide connections with the outpatients building at low levels and also preparing for the future rebuilding of the outpatients building.

总建筑面积	51,849 ㎡
设计期间	1997/06~2000/06
建设期间	2000/12~2003/06
竣工年月	2003/06
配合设计	中岛建筑事务所

摄影：SS名古屋株式会社

日本岐阜县综合医疗中心
Gifu Prefectural General Medical Center, Japan

总建筑面积	57,950 ㎡
设计期间	2000/10~2003/01
建设期间	2003/12~20011/02
竣工年月	2009/12
配合设计	中岛建筑事务所

作　　为担当广泛领域的急性医疗任务的基础医院，在不缩小、不中断现行功能的基础上实施重建和改建计划。在有限的建筑用地内最大限度地缩小临时建筑的规模，控制建设成本。

医院同时作为急救中心，设置直升机停机坪，采用抗震结构等防灾措施。在整个建筑内部，患者与职员动线分离，病房楼L形的主要部分内集中布置服务功能和重症病床。同时还积极推行太阳能发电，废热利用等环保节能措施。

This was a refurbishment and upgrading project for a main hospital that handles acute medical treatment in a wide range of fields, carried out without any reduction or stoppage of existing functions. Temporary works on the restricted building site were reduced to a minimum, reducing the construction costs.

This hospital is also a base for disaster response, so a heliport was installed, and other measures against disasters were taken such as base isolation structures, etc. In the building as a whole, the circulation of patients and staff were separated, and in the hospital building the service functions and the beds for serious cases were concentrated in the main parts of the L-shaped building. Also, environmental measures were taken such as solar power generation and use of waste heat, etc.

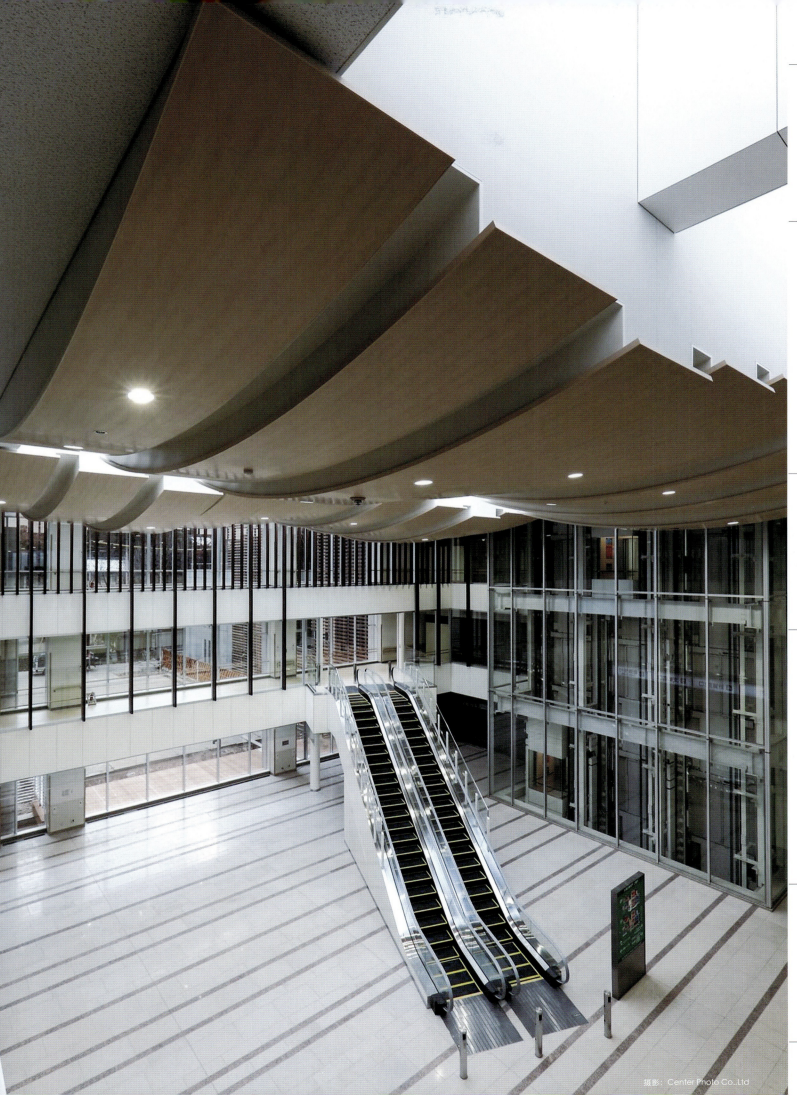

| 174 | 三亚时代海岸项目
Sanya Time Coast Project

| 176 | 越南突敦克项目
Thu Duc House Hiep Phu High-Grade Apartment Building,Vietnam

| 178 | 迪拜英加斯综合开发项目
INJAZ District Development,Dubai

| 182 | 日本日暮里花园城
Sunmark City Nippori , Japan

Residential Facilities
住宅项目

三亚时代海岸项目
Sanya Time Coast Project

总建筑面积	88,250 ㎡
设计期间	2008~2009
建设期间	2009~2010
竣工年月	2009/10
摄影	日本设计

三亚时代海岸项目位于海南省三亚市河口区港门下村，与南海相连的临春河河口相连，南侧与鹿回头山隔海相望。用地呈不规则五边形状，是融酒店、酒店式公寓、商业、游艇俱乐部四大功能的综合性城市建筑项目，设计充分利用场地的地理优越性，以"梦之海岛"为整体设计理念，以"海"为设计元素创造风格独特的街区。协调"水、风、光"自然要素，创造风格新颖的街区。

总平面布置综合考虑与原有建筑的协调统一性，充分利用山、海等自然景观。

The Sanya Time Coast Project is located in the southern part of Gangmenxia Village, Hekou District, Sanya City, Hainan Province. The site is an irregular pentagon, an integrated urban construction project with the four functions of hotels, apartments, commerce and yacht club, but also belongs to the urban core architecture group which must be designed together as a holistic urban environment. The project site is located in the estuary of Linchun River which flows into the South China Sea; from its south, the Luhuitou Mountain can be seen across the sea. The project makes full advantage of the superb geographical attributes of the site, employing "Dream Island" as the overall design concept to create unique neighborhoods with the "sea" as design element. It is a new approach to block design that coordinates natural elements of like "water, wind, and light".

The general layout will comprehensively consider coordination and unity with the original building, making full use of the mountain, sea and other natural landscape.

越南突敦克项目
Thu Duc House Hiep Phu High-Grade Apartment Building, Vietnam

越南胡志明市的高级住宅项目。业主守德住宅开发公司是一家开发和销售中档公寓的房地产企业。而本项目作为该公司开发的第一个高级公寓项目，代表着该公司崭新的形象。本项目位于胡志明守德九区，该地区规划有高速公路和地铁，具有很大的发展潜力。

有两点需要在此一提，一是在基地内实现了完全的人车分离（除消防车等特殊车辆），二是在幼儿游乐场地与公共空间内配置了极为安全可靠的安保管理系统。343个住户分布于二层至二十层。为了提供更好的通风条件，在八至二十层的"巨大墙面"中设有大开口，利用这一巨大的开口，满足规范要求的变形缝。

本建筑的外形独具匠心。首先在与高速相邻接的位置条件中体现出了"轻快"之意，随后又通过圆弧状的线性平面计划进一步延伸出了"速度"与"流向"这两个概念。此外还利用空气的流向，在风水上佳的位置打通一个大的通风口，最后融合了其他的一些手段，最终决定出了本建筑物的设计方案。

关于住宅层，充分考虑到风的流通性进行了设计。通过单廊基准的住户配置、金属线外皮、巨大的通风口等，将公共部分空间的空调区配置降低到了最小限度。

总建筑面积	69,020 ㎡
设计期间	2008/4~
建设期间	2010/2~2011/12(预计)
竣工年月	2011/12(预计)

Thu Duc House Hiep Phu High-Grade Apartment Building is an up-market housing project in Ho Chi Minh City of Vietnam. The owners Shoude Housing Development Corporation is a real estate company which mainly develops and sells mid-range apartments in the area centered in the 9th District of Ho Chi Minh. It is the first up-market apartment project developed by this company and represents the company's new image. Highway and subway are planned for the district creating great potential for future development.

Two points need to be mentioned here. The first is that a complete separation of people and vehicles at ground level is achieved (except for fire engines and other special vehicles); the second is that a highly secure and reliable security management system is provided for the child playground and public spaces. 343 housing units are distributed from the 2nd floor to the 20th floor. Large openings are set in the "great wall" from the 8th floor to the 20th floor to provide better ventilation.

The large opening is specified to meet the requirements of deformation joints. The ceiling height of the 2nd floor and the 20th floor are raised for the up market units with four beds on these stories. The shape of the building is distinctive. First, the "light" is reflected in the location adjacent to highway. This is then extended further by arc-shaped linear flat plans visualizing the two concepts of "speed" and "flow". In addition, a large vent is opened in the optimum location for flow of "Chi" according to fengshui. Finally, everything else is integrated to determine the final design scheme of the building.

On the residential floor, the design fully integrates the ventilation. The air-conditioned area of public space is reduced to the minimum by the housing unit configuration, metal frame grid, huge vents, etc. of single corridor baseline and constant improvement.

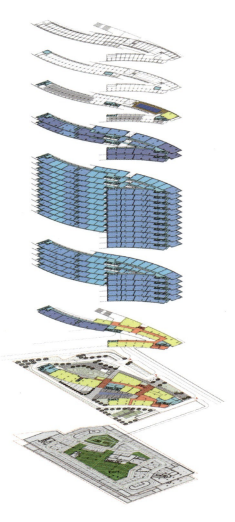

迪拜英加斯综合开发项目
INJAZ District Development, Dubai

| 阿 | 联酋迪拜的高层综合住宅项目概念设计。用地位于世界第一高塔Burj Dubai旁的高级住宅区内。在设计方面，业主提出：设计不仅要体现高品质，还要成为该开发区的象征。我们的设计团队在满足以上条件的同时 积极地将现代元素与阿拉伯的传统意向不断融合到设计当中，致力于创作最为优秀的作品。 |

项目由5座高层塔楼构成。塔楼标准层的形状独特，是由4个正方形呈雁行状组合而成，从而可以布置更多的视野开阔的房型，为住户提供种类更为丰富的住居方案。此外，为使外观更具特点，还在各塔楼的顶部建造了一个玻璃立方体，人称"灯笼"。在这些玻璃立方体内，设有居住者专用的泳池、健身房、SPA及休息区等空间，极大程度地提高了居住空间的舒适性。到了晚上，这些灯笼一同亮起，形成一条色彩独特还极富韵律感的天际线，悬浮在高空中营造出一道极具特点的风景线。

地上停车场上方仿照阿拉伯传统街区"Madinat"建造了各式各样的庭园与小径。富有人性尺度的空间穿插于塔楼与塔楼之间，咖啡厅以及商店被随机地设置在其中。这种新颖而又独特的设计方式，通过为市民提供空间起到了加强地域间人们的交流沟通的作用。

总建筑面积	151,910㎡（地上）
设计期间	2008/4~2008/10（概念设计）

INJAZ District Development is the concept design of high-rise integrated residential projects in Dubai, UAE. The plot is located around Burj Khalifa tower--- the highest tower in the world, within the planned senior residential area. The owners proposed that the design should not only reflect high quality but also become a symbol of the zone. The design team accepted the challenging conditions and actively integrated modern elements with traditional Arab motifs into the design, and is committed to creating an outstanding work.

This project is composed of 5 high-rise towers. The shape of the typical floor plate of the tower is unique, consisting of four wild goose-shaped squares. This allows more offices with broad vistas. It becomes possible to provide tenants with more variation in residential types. In addition, a glass cube called the "lantern" was placed on the top of each tower for a distinctive appearance. In the glass cubes, there are a swimming pool, gym, SPA and rest areas and other spaces only for tenant use, which greatly improve the amenity level of the residences. In the evening, all the lanterns are lit at the same time, producing a very rhythmic skyline with unique color, suspended in the air to create an extremely unique silhouette.

The parking is above ground. An artificial ground is built over the entire parking lot. On the artificial ground, a wide variety of gardens and trails was built by imitating the traditional Arab "Madinat" blocks. The space is designed to a human scale and interspersed among the towers. Cafes and shops are set among them randomly. This novel and unique design promotes communication among people from different functions by providing space for the public. Notes on environmental techniques:

日本日暮里花园城
Sunmark City Nippori, Japan

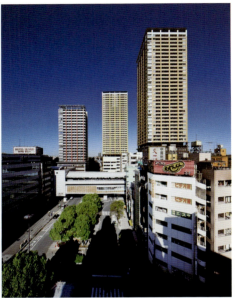

总建筑面积	22,255㎡, 52,801㎡, 42,595㎡
设计期间	2000/04~2002/03
建设期间	2005/01~2007/05
竣工年月	2007/05
摄影	SS东京株式会社

日暮里站前地区再开发的目的是通过再开发紧密连接站前的3个区域，把日暮里站前地区打造成为洋溢着生机与活力，商业繁华的城市空间。将原先城市的老旧房屋改变成具有时代感的超高层建筑，形成了东京北部的门户。日暮里车站同时是连接成田机场的门户，包括京城机场线、JR山手线等交通都在日暮里汇集。

这3个区域的建筑都是由高层居住空间、低层商业及办公空间构成的综合设施，并在每栋建筑的三层设置了与日暮里车站相连的步行平台，实现了站前3个区域与日暮里车站和即将开通的新交通系统的衔接，形成行人的立体回游空间，增加了站前地区的活力和交通便捷度，对站前地区的无障碍化也起到了积极的作用。

邻接的3栋建筑采用统一的设计风格，但是要求分阶段施工，并与即将开通的新交通系统的站房、步行平台相连，还需要同时推进日暮里站的改建工程及站前广场的改建规划等，整个再开发项目呈现出非常繁杂的特点。针对这一特征，"日本设计"在每个区域配备了一个设计、监理团队，并全程提供顾问服务，使设计团队之间能够紧密联系，从而顺利推进这一复杂的项目。整个设计方案注重日暮里的地区特色，建筑高层外墙采用源自传统地名"日暮里"的日暮时天空的颜色，每栋建筑采用日暮时天空的不同颜色，3个区域合为一体就会呈现出日暮时分天空的色彩组合。裙房部分则都采用象征大地的茶色系，演绎出沐浴在夕阳余晖中连绵的谷中山丘。

This project's intention was to combine 3 "Higurashi-no-sato" areas and to redevelop them to bring back vitality and dynamism. Each area consists of mixed use complexes with high-rise apartments and commercial and work spaces on lower floors.

Each complex is connected on the 3rd level to the Nippori train station by a pedestrian deck. The complexes are connected to the new circulation roads. There is also a multi-level dedicated pedestrian space on the ground level promoting life and convenience, designed to be a barrier-free universal design.

The project, although planned on 3 different timelines is built as one whole unified and connected project. The project required a very complicated coordination involving many other projects such as the connection to the new circulation road, train station, pedestrian deck, the construction works of Nippori Station Renovation, and the plaza in front of the station. For this project, Nihon Sekkei had to assign separate staff to form design and supervision teams in each of the areas. The company also took on the role of project consultant to deal with the complexities of the project execution that demanded close communication between all of the teams. With this coordination, it was then possible to provide a unified design concept that reflected the character of Nippori for each of the areas. The color of the buildings reflects the name Nippori (meaning village of sunset), adopting the colors of the "sky at Sunset", and adopting the traditional name of Higurashi-no-sato. Each building was not finished in one particular color, but rather a gradation of colors across the building, mimicking the sunset. Also, the lower portion was colored in an earthy color to mimic the ground to project the image concept of "a hill within the valley".

186	无锡市综合交通枢纽	
	Wuxi Comprehensive Transportation Hub	

190	上海马陆项目方案设计	
	MALU Project, Shanghai	

194	广州珠江新城核心区地下空间及中央广场建筑设计	
	The Competietion of the Underground Space and Central Square of Pearl New Town, Guangzhou	

200	日本品川车站综合体规划及开发	
	Shinagawa Station Development Project, Japan	

Urban Transit Complex

城市交通综合体

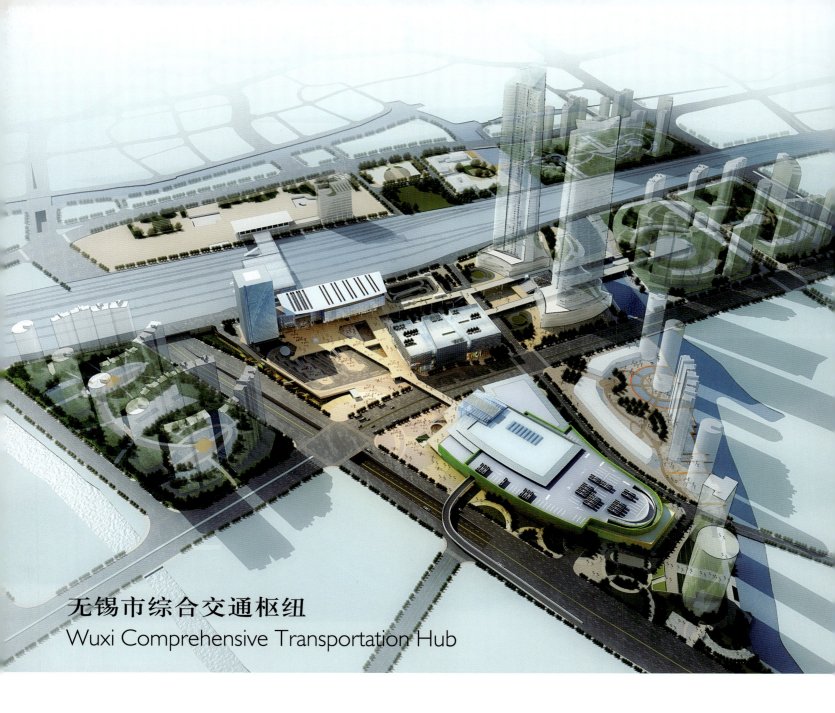

无锡市综合交通枢纽
Wuxi Comprehensive Transportation Hub

总建筑面积	248,119㎡（B2区）124,294㎡（F2区）
设计期间	2008/11~2009/9
建设期间	2009/6
竣工年月	2011/12
配合设计	浙江大学建筑设计研究院
	无锡市建筑设计研究院
摄影	林铭述摄影工作室
获奖经历	国际设计竞赛一等奖

无锡，作为地处长三角地区经济和工商业中心的城市，伴随着交通大动脉沪宁城际铁路2010年的开通，期待着为城市经济的持续高速发展带来新的活力。而未来将成为无锡大门的火车站北广场地块内将建成两条轨道交通的车站，会同规划于此的长途汽车站，促使交通功能进一步得到充实完善，变得更为合理。实现沪宁城际铁路、原有普速铁路以及长途汽车等组成的对外交通和由地铁1号线、3号线、近郊公交、市内公交、出租车、社会车辆组成的市内交通之间的综合性"零换乘"，也同时成为高科技无锡的崭新门户，代表着无锡的现代化新形象。

解决现状规划的课题，将打造无锡新大门、创造城市新形象，实现零换乘的公共交通和流畅的车辆交通以及利用北广场连接南北两侧的城市功能作为主要概念，完善本区域路网并规划设计主要功能和流线。

建筑设计的用地分为两个部分，沿用规划阶段的编号分别为面积76136m²的B2地块和面积64868m²的F2地块。

建筑设计在深化规划阶段的概念和完善规划成果的同时，具体在解决与火车站南广场的联动、与城际铁路和地铁等轨道交通的接续、处理与周边道路以及周边用地的结合

等课题方面进行了深入推敲,并确立以下的设计理念:
1. 功能上追求快捷高效,强调作为交通设施的便利性。
2. 重视作为无锡市门面的建筑地标性,结合交通设施特点,确立了经典、简洁、现代的外观设计理念。
3. 在不同标高和平面位置营造丰富的广场空间,方便使用者,体现以人为本的主导思想。
4. 积极利用架空屋面和屋顶绿化,打造环保型建筑,实现发展的可持续性。

B2地块的总体布局方面,在城际铁路的进出站口位置设置交通综合体,利用大量上下车人流从中穿过这一优势,引入商业和办公功能,根据就近使用的原则,方便通过人流的同时,也有效地缓解用地内外的交通负荷。在用地西北角处设置商业设施,与交通综合体内的商业设施形成呼应,相互强化集客效果的同时,避免了对众多交通设施的干扰。

除此之外,在用地东侧地面设置市内公交枢纽,将路面公交系统与城铁、地铁、长途汽车以及现有铁路等其他交通手段有效结合。用地的西南角靠近交通综合体位置设置出租车上下点,利用广勤路处理必然会出现在火车站前的大量出租车和社会车辆。

F2地块以长途汽车站为主,利用火车站前交通设施密集的优势在南端实现与各交通系统的连接。在F2地块的北端北新河一侧设置配套住宅,为大面积的公共设施提供辅助性支持。

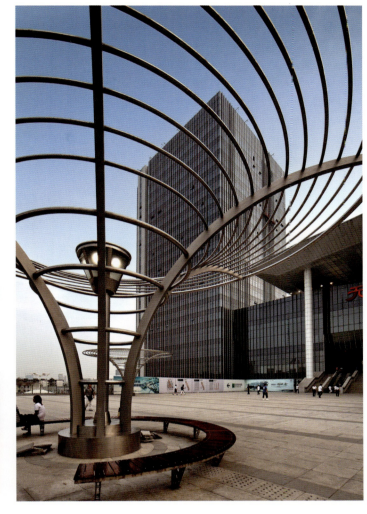

Wuxi, a city as the economic, industrial and commercial center located in the Yangtze River Delta region, expects to bring new vitality for the continued rapid economic development with the opening of Shanghai-Nanjing Intercity Railway as a traffic artery in 2010. In the North Square block of Railway Station which will become the door of Wuxi in the future, two rail stations will be built. The coach station planned here with it will promote transport function to be further improved and become more reasonable. A comprehensive "zero transfer" in the external traffic such as Shanghai-Nanjing Intercity Railway, the original normal speed railway and coach and urban transport vehicles such as Metro Line 1, Line 3, suburban bus, city bus, taxi and social vehicles can be achieved; meanwhile, it will also become a brand new portal for high-tech Wuxi and represent the new modern image of Wuxi.

The main concepts are: to solve the status quo planning, create a new door in Wuxi and a new city image, achieve zero transfer public transport and smooth vehicular traffic, connect north and south by use of the North Plaza; improve the regional road network; plan and design the main functions and flow lines.

The architectural design land is divided into two parts: B2 plot with an area of 76,136 square meters and F2 plot with area of 64,868 square meters in the planning stages When the building design is deepened and planning outcomes are improved, in-depth scrutiny is done in the subjects such as the linkage with the South Railway Station Square, connection between inter-city railway and metro or such rail transportation, the connection between the surrounding roads and the surrounding lands. And the following design philosophies are established:
1. In function, we pursue speed and efficiency, emphasizing the convenience of transport facilities.
2. We place emphasis on the landmark of Wuxi City; combining the features of transportation facilities, we establish the classic, simple and modern design concept.
3. In different elevations and level locations, we create a rich square space, which is convenient for users. This reflects the guiding philosophy of people-oriented.
4. By active use of overhead roof and green roof, we build environment friendly buildings to achieve sustainable development.

As for the overall layout of plot B2, transportation complex is set in the entrance and exit of intercity railway station; by taking the advantages of a large number of people through it, we can introduce commercial and office functions. According to the principle of using at the nearest place, it is convenient for stream of population; meanwhile, it can also effectively alleviate the traffic load of inside and outside land.

Commercial facilities are set at the northwest corner of the land and echo with the commercial facilities in the integrated transportation. When the population gathering effect is mutually reinforced, interferences on the numerous transport facilities are avoided.

In addition, the city transportation hub is set in the east land, which connects the road transportation system with the suburban railway, metro, coach and existing railway and other transport means effectively. The southwest corner of land is near the taxi station in the complex traffic. By using Guangqing Road, it is bound to produce a large number of taxis and social vehicles in front of train station.

F2 block is mainly designed as coach station. By taking the advantage of intensive transport facilities in front of train station, realize the connection with various transport systems in the south. Supporting residence area is set in the northern end of plot F2 – the North New River to provide ancillary support for large areas of public facilities.

上海马陆项目方案设计
MALU Project, Shanghai

总建筑面积	189,460㎡
设计期间	2008/6
建设期间	2008/12
竣工年月	2010/11(一部)
配合设计	汉嘉设计集团股份有限公司

一、基于广域城市开发的视角所开展的站前开发

马陆站前开发是以马陆站为中心的复合型大规模开发项目，也是决定今后马陆站周边开发方向的重要工程项目。因此在确定本规划方案的设计方向时，不仅要充分考虑本地块的开发定位，还需要充分把握马陆站周边的广域城市开发的未来定位，这一点极其重要。由此我们的设计理念之一定义为"从广域城市开发的视角开展马陆站前开发"。

二、资源节约型·环境友好型站前开发

本规划方案着重考虑被视为当今世界重大的社会问题之一的环境问题。为了创造"资源节约型·环境友好型的站前城市空间"，在设计中我们尽力控制建设及维护所需的能源，减少其对周边城市环境以及

人们的城市生活所带来的不利影响。

上述的"基于广域城市环境的视角·地球环境的视角所展开的站前开发"是"日本设计"最为擅长的领域。21世纪被称为环境的世纪，我们将发挥多年来积累的设计经验，全力打造适合于21世纪的新颖独特的站前开发设计方案。

三．设计方针
1. 建设促进马陆广域性发展的城市轴，整个用地以车站为中心一体化设计。

  在马陆站周边地区的上位规划中，预计将在沿着富蕴南路向南北方向伸展的城市轴线上，集结商业、办公功能。同时，马陆站南侧是外环西沿线，因此从城市发展趋势来看，马陆站北侧将成为城市发展的中心。

  住宅区规划设计为整体面积约为8.1万㎡，约有900户的住宅和其附属设施。利用公共空间将车站、商业、办公设施以及住宅合理地联系，创造出便捷、舒适的城市环境，将一体化城市功能的优点最大限度地发挥。地面和屋顶绿化采用统一的环境设计手法，将不同功能有机地联系，以强调整体的设计风格，打造崭新的自然共生型街区。

2. 强化车站的交通枢纽功能，站前空间彻底实现人车分离。

在轻轨站的周边集中设置道路、公交汽车站、出租车等候区，使人们能够便捷地来往于各个设施之间，最大限度地强化马陆站的交通枢纽功能。同时设置步行平台，连接位于地面7.5m高处的马陆站中央大厅和轻轨车站、公交汽车站、出租车等候区，实现人车动线的立体化分离。

3. 实现舒适的，人与环境共生的"绿色山丘"。

车站北侧建设的马陆中心圆形广场是一个展示街区新形象的空间。横穿圆形广场大屋顶、连接屋顶花园的绿化大阶梯、形状独特的标志性塔楼、伸向周边道路的绿化散步道共同描绘出环境型站前广场的城市景观。

我们在"从广域城市开发和地球环境的视角进行的站前广场开发"这一先进的设计理念指导下，致力于打造"人与环境共生的未来型城市环境"，并希望通过将马陆车站周边地区建设为一个"成为中国未来城市开发典范的城市环境"。

Ⅰ. Based on the development before station from the perspective of broad urban development

Malu pre-station development is not only a complex large-scale development project centered as Mulu station, but also an important project to determine the future direction of the surrounding areas of Malu stations. Therefore, when determining the design direction of this scheme, we not only give full consideration to the development and position of the block, but also need to master the future orientation of broad urban development of the surrounding areas of Malu stations. It is extremely important. One of our design philosophies is defined as "Based on the development before station from the perspective of broad urban development".

Ⅱ. Resource-saving and environment-friendly pre-station development

The program focuses on environmental problem which is deemed as one of the major social problems in the current world. In order to create "resource-saving and environment-friendly pre-station urban space", we try to control the energy required to construction and maintenance in design, reduce the adverse effect on the surrounding urban environment and urban life.

The above "Based on the development before station from the perspective of broad urban development" is one of the most expertise areas of Japanese Design. The 21st century is called the century of environment. We will give full play to the accumulated experience in design over the years, to create an innovative and unique pre-station development scheme which is suitable in the 21st century.

Ⅲ. Design approaches

1. Build the urban axis to promote Malu broad development / the whole land is integrated and designed with station as the center.

In the upper planning of the Malu surrounding areas, it is expected that business and office functions are gathered together on the city axis stretched along the South Fuyun Road in the north-south direction. Meanwhile, the south of Malu station is the western line of outer ring. Therefore, the north side of Malu station will become the center of urban development from the trend of urban development.

The residential area is set with overall area of about 81,000 square meters and about 900 houses and their ancillary facilities. By using public space, we connect station, commerce, office facilities and residence reasonably to create a convenient and comfortable urban environment; the benefits of integrated urban functions can be played to the maximum. The ground and roof vegetation adopt unified environment design approach and link different functions organically, in order to emphasize the overall design style and create a new block of natural symbiotic type.

2. Strengthen the hub function of the station / full realization of people and vehicle before the station

Around the light rail station, we set roads, bus stops, taxi waiting area, so that people can easily travel between various facilities, the transportation hub function of Malu Station can be maximized. At the same time, we will set foot platforms, connecting the central hall and light rail station, bus stops, taxi waiting areas at the height of 7.5m above ground; the three-dimensional separation of people and vehicles lines can be achieved.

3. To achieve the comfortable "green hills" where human and environment are in symbiosis

The circular plaza in Malu center will be built in the north of the station, which is a space for displaying new image of block. Cross the great roof of Circular Plaza, a large vegetation ladder connecting roof garden, unique iconic tower, the vegetation pedestrian which stretches towards the surrounding roads, depicts the environment urban landscape of pre-station square.

Under the guidance of the design concept of "pre-station square development from the perspective of broad city development and global environment", we will devote ourselves to create "the future urban environment where people and environment are in symbiotic" and hope to build the surrounding area of Malu station as an "exemplary urban environment for China's future urban development".

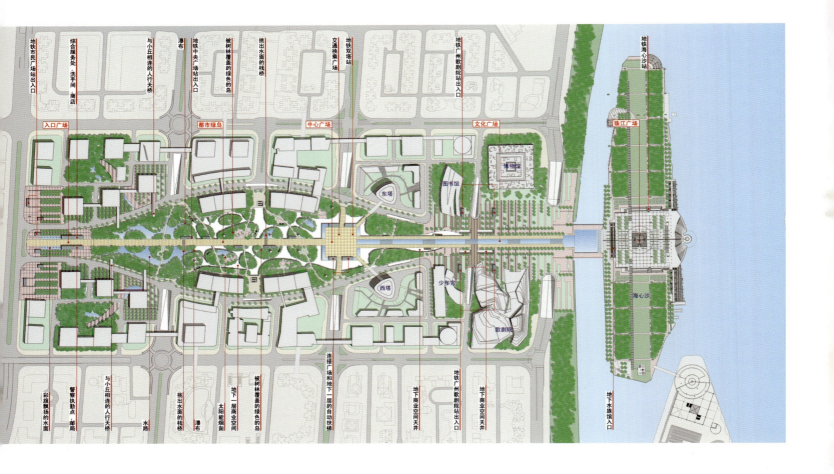

广州珠江新城核心区地下空间及中央广场建筑设计
The Competietion Of The Underground Space And Central Square Of Pearl New Town, Guangzhou

设计期间	2005/10~2006/10(竞赛) 2006/12(深化)
合作设计	广州市城市规划勘测设计研究院
	广州地下铁建筑设计有限公司
获奖经历	国际设计竞赛第二名

珠江新城核心区地下空间和中央广场位于珠江新城21世纪CBD的主轴线之上，由地下交通设施、商业设施、中央广场组成的综合标志性城市中心设施。我们的设计提案将解决下列6个方面的问题。

1. 象征性

作为广州城市新中轴线的核心区，需要解决城市景观象征性问题，以珠江之水为主题，寓意广州珠江三角洲的城市特色，塑造广州山水城市的标志性。

2. 便捷性

通过改善珠江新城现有路网，形成完整的地上地下一体的公交系统、道路系统，改善商务中心区的交通组织，构筑地下集运系统、地下道路系统、地下交通枢纽三位一体的立体地下交通体系，强调灵活性、联动性、便捷性，与周边区域形成联通网络。

3. 安全性

地下空间的防灾安全问题，建设安全第一的、开放式的、明快的地下空间和交通系统，着重解决防灾、人员疏散、对抗突发事件等，做到面面俱到，万无一失。地下一层是直接对外开放空间，实际上

是扩展的避难空间。消防车可以沿着下沉街道直接驶入，结合各处的避难楼梯、电梯实现救援消防活动。

4. 舒适性

考虑广州岭南亚热带气候特色，解决地下空间的节能问题，体现自然中的多元活动空间，与绿色相应的交通设施，围合出舒适的地下空间环境。

5. 先进性

建设宜人的地下空间，解决与周边区域功能互补的问题，积极采用节能技术，合理体现环保节能理念。利用浮岛剖面特点，引入自然通风，采光，从绿化、水、光、风等产生的可再生能源。

6. 经济性

重视实施性和经济方面的问题，特别是解决结构和施工的复杂性问题，注重可实施性，优化施工方法，借鉴成熟的地下空间开发经验，建设可持续性发展的地下空间和中央广场。

整个地下空间和中央广场从南到北分为5个部分：

入口广场——地铁市民广场站

采用几何学外形，通过旗帜、水体、玻璃构筑物、树木形成具象征性的入口空间。

都市绿岛——地铁中央广场车站

行人可回游通过桥梁连接的地表的"绿色岛屿"，"流动"到地下一层的室外购物中心与文化活动。

中心广场——城市枢纽 地铁双塔站

边长88m的都市中心广场相当于海心沙市民广场的一半，丙者在广州市新中心轴线上形成稳定的规划布局。北侧商业区域与南侧商业区域通过扶梯经由中心广场，跨越花城大道。从地下过渡到地面的行人可看到中心广场的开放感。它既是景观平台，也是标志性的纪念广场，同时还是整个交通系统的枢纽中心。

文化广场——地铁广州歌剧院站

珠江广场——海心沙岛——地铁海心沙站

Underground space and the center of Pearl River New City Plaza, Pearl River New City in the 21st century in the CBD on the main axis, from the ground transportation facilities, commercial facilities, an integrated Central Plaza City Centre landmark facilities. Our design proposal will address the following six aspects.

1. Symbolism

As the core area of the new axis of Guangzhou city, we need to address the problem of symbolism for urban landscape; the Pearl River is the theme implying the city features of Guangzhou Pearl River Delta, and we will create a symbol for the landscape city of Guangzhou.

2. Convenience

By improving the existing road network of Pearl River New City, we will have a complete integrated aboveground and underground public transportation system and road system, improve the traffic organization in the Central Business District, and build the three-dimensional underground transport system consisting of underground gathering system, underground road system and underground transport hub. Place emphasis on flexibility, interaction and convenience;

connected network is formed with its surrounding area.

3. Security

Disaster prevention and safety issues of underground space: build safety-first, open, bright underground space and transportation system with emphasis on disaster prevention, evacuation, dealing with unexpected events. We should prepare for all of these situations without fault. The 1st floor underground is an open space directly to the outside; in fact, it is an extended shelter space. Fire engine can directly go into along the sunken street; combined with escape stairs, elevators can achieve the functions of fire rescuing.

4. Comfort

The features of sub-tropical climate n Guangzhou Lingnan is considered; solve the energy-saving problems of underground space, reflect the diverse activity space in nature, and transport facilities corresponding with green, create a comfortable and enclosed underground space environment.

5. Advancement

Build pleasant underground space, solve the problem of complementary function with surrounding areas, actively use energy-saving technologies, reasonably embody the idea of environment protection and energy saving. By using the profile characteristics of floating island, we can introduce natural ventilation and lighting, renewable energy from the vegetation, water, light, wind etc.

6. Economy

Place great emphasis on implementation and economy, in particular address the complexity problems of structure and construction; attention should be given to the implementation of causes; optimize construction methods; draw on mature experience in development of underground space; build underground space and Central Square with sustainable development.

The entire underground space and central square are divided into five parts from south to north:

Entrance Plaza — metro station in civic square
By using geometry shape, we can form a symbolic entrance space together with banner, water, glass structures and trees.

Urban Green Island — metro station in Central Square
Pedestrians can go to the outdoor shopping center and cultural activities on the 1st floor underground by "Vegetation Island" which connects bridge with ground surface.

Central Square — urban hub, metro station in twin towers
The urban center square with 88 meters side length is equivalent to the half of Haixinsha Civic Square; both of them form a stable layout on the new central axis of Guangzhou City. The northern and the southern business districts extend through escalator, cross the Flower City Avenue. The pedestrian from underground to aboveground can see the open feeling of Central Square.
It is a landscape platform but also a symbolic commemorate square, and also is the hub of the entire transportation system.

Cultural Plaza — Metro Station in Guangzhou Opera House

Pearl Plaza — Haixinsha Island — Metro Station in Haixinsha

摄影：川澄建筑摄影事务所

日本品川车站综合体规划及开发
Shinagawa Station Development Project, Japan

总建筑面积	337,216㎡
设计期间	1987/6~1995/5
建设期间	1995/6~1998/11
竣工年月	1998/12
获奖经历	建筑业协会奖、优秀消防防灾系统、
	空调·卫生工学会奖 建筑设备部门·技术奖、
	电气设备学会奖 技术部门·设施奖、
	日本建筑学会业绩奖

在不断变化的城市风景中，如何使"品川国际城"保持恒久的魅力，是设计的主要课题。本案由三栋超高层建筑组成，建筑与建筑之间的中庭空间与建筑本身共同形成建筑群，这样的建筑群将为塑造城市形象做出积极贡献。在城市中营造中庭空间，保障光线与清风在城市中的流转。

如何通过建筑反映自然风景是外立面设计的主题。超高层外立面采用预制混凝土板贴水磨天然石材，外包玻璃，通过内部凹凸不平的混凝土板反射出一天中不断变化表情的阳光。近景方面，透过玻璃还可以欣赏到水磨天然石材的柔和肌理，从而打造出令人耳目一新的超高层建筑。沿空中走廊布置中庭、画廊，形成在水平垂直方向连绵延续的精彩空间。这些空间在打造视觉上的通透感的同时，还通过阴影使人感受到一天中光线的变化。透明的外表皮与形成骨骼的框架交织在一起，为空间赋予流动感和生命力。

撮影：Miyagawa Co.,Ltd

摄影：川澄建筑摄影事务所

The theme of the project was how "Shinagawa Intercity" could maintain its attraction within the changing urban landscape. This project consisted of 3 high-rise buildings. We were strongly focused on the contribution of this group of buildings to the city be not only in the actual form of the buildings, but also in the void spaces created between the buildings. We were strongly aware of the urban spaces through which light and wind can gently pass. Achievement of such spaces was enabled by this integrated project, and an attractive urban landscape was formed by including the adjoining pedestrian spaces.

The theme of the façade design was how to reflect nature. This scheme consisted of high-rise buildings having PC façades with exposed natural stone. However, these were enclosed in thin glass in which the sky is reflected, and through which the changes in the sun throughout the day are reflected in the deep depressions and protrusions in the PC façade. The gentle texture of the exposed stone that can be seen through the glass close up gives an image that is unlike that of any other high-rise building.

Also an internal and external space referred to as the atrium gallery is provided along the skyway, with a spatial composition that is continuous in both vertical and horizontal directions. These spaces have a composition with a transparent feeling that is easily confirmed visually. At the same time by creating shadows the changes of the sun during the day can be felt, and the transparent skin and its frame appear layered, so that the theme of the design is the creation of flow within the space.

206 迪拜城规划
Bawadi City Planning, Dubai

209 迪拜海岸规划
Palm Deira Master Plan, Dubai

210 上海陆家嘴金融中心区都市设计
Lujiazui CBD Urban Planning, Shanghai

214 苏州工业园区都市设计
Suzhou Industrial Area Urban Planning

218 潍坊安顺新区整体规划及奥林匹克中心建筑概念设计
Olympic Center Conceptual Architecture Design and Anshun New Town Urban Planning, Weifang

222 长春高新开发区城市设计
Hi-Tech Center Urban Planning, Changchun

226 威海金线顶城市规划
Jinxianding Urban Planning, Weihai

Urban Planning Projects
城市核心空间规划

迪拜城规划
Bawadi City Planning, Dubai

总建筑面积	818,239 ㎡
建设期间	2008/5~2008/8

迪拜城总体规划的目标：
- 通过发展世界一流的食宿、娱乐和购物天堂，创建独一无二的城市设计。
- 创建面向国际的住宅发展区域。
- 创建能够反映世界公认最佳实践原则的可持续发展环境。

目标
- 完成有益于城市发展的高品质城市设计，此设计能够反映出作为迪拜园核心吸引力的迪拜城的地位。
- 通过建筑设计、概念图、引导标志和种植花木来强化建筑环境的特征，进而营造强烈的社区感。
- 在公共领域设计期间，通过建筑形式、街景和其他公共区域设计具有吸引力的街景。
- 鼓励采用气候敏感性设计，及在公共和私人领域内采用可持续发展的技术和方法。
- 强化Bawadi Boulevard区内以及毗邻Bawadi Boulevard区的街道特征，并且在整个城市发展中提高行人的渗透率，鼓励人们采用步行方式外出。

The aims of the Bawadi Master Planning:
- Create a unique urban design through developing a world-class hospitality, entertainment, and shopping destination.
- Create a unique residential development for international use.
- Create a sustainable environment reflecting recognized international principles of best practice.

OBJECTIVES
- To achieve high-quality urban design for development that reflects the role of Bawadi as a core attraction within Dubailand.
- To create a strong sense of community by enhancing the character of the built environment through building design, massing, signage, and planting.
- To develop an attractive streetscape during the design of the public realm through the built form, streetscapes and other public areas.
- To encourage climatically sensitive design and the use of sustainable technologies and practices within the public and private realm.
- To strengthen the character of the streets within and adjoining the Bawadi Boulevard district and enhance pedestrian permeability throughout the development in order to encourage pedestrian movement.

| 总建筑面积 | 46,000,000 ㎡ |
| 建设期间 | 2004/9-2005/9 |

迪拜海岸规划
Palm Deira Master Plan, Dubai

本案同时拥有大规模商务区、阿拉伯传统帆船港等海上交易中心，是城市化程度更高的人工岛。

人工岛的形状模仿当地人认为能够为游客带来好运的吉祥树的形状，拥有狭长人工沙滩的岛屿被环礁状岛屿环抱。环礁作为内岛的防波堤，保护其不受海浪冲击。尽量不破坏阿拉伯湾潮汐对溪流生态系统和自然水系的净化作用，并由此确定距离市中心最近的被称为"根系"的地域形状定。如何维持整个项目的水质也是规划设计需要解决的重要课题。大量采用通过潮水涨落促进水质净化的运河结构。另外，由于老城区难以提供足够的水、电和热源，因此需要建设热电联供、地区冷气、海水淡化设备等成套设备网络。

This project was for an urban man-made island having a large scale business zone, together with a marine trading center such as a port for traditional Arab dhow ships, etc.

The man-made island has long artificial beaches that incorporate reef islands, and is in the shape of a "palm tree", a tree that is considered to be an omen of good luck to travelers in this region. These reefs have the role of wave break to protect the island on the inside from wave damage.
In this project close attention was paid to ensure that there would be no damage to the ecosystem of the creeks and to the natural purification action of the waters as a result of tidal movement in the Arabian Gulf. The shape of the area closest to the city, which is known as the "root", was determined taking this into account. Maintenance of water quality was the most important issue in the project as a whole. Many schemes such as canals, etc., were incorporated to promote the water purification action of the ebb and flow of the tides. Also, it was not possible to rely on the old city to provide the necessary electrical power, heat sources, and water, so a plant network that included co-generation, district cooling, and desalination facilities, etc., was provided.

上海陆家嘴金融中心区城市设计
Lujiazui CBD Urban Planning, Shanghai

上海浦东陆家嘴地区作为上海高速发展的代表和世界著名旅游景区，伴随着金融街的拓展，我们力求打造区别于以往大型街区的工作与生活并重的人性化街区的立体解决方案。

为此，我们制定出如下三大设计目标：

1．具有高附加值的核心区：形成金融机构、房地产、投资、咨询等跨国企业云集的具有高附加值的核心区。

2．交通便捷的商务区：确保机动车便捷地出入地块和停车。完善作为公共交通的轨道交通、公共汽车以及前往各地块的行人动线。充分利用地下空间，形成短时间内迅速分散人流的行人网络。

3．高品质的综合功能区：完善会展、文化、酒店等功能，为金融中心区提供高品质的配套。充实居住功能，全面提升综合功能区的繁华和魅力，将建筑裙房、地块内的开放空间连接成网络，在多姿多彩的繁华空间和舒适的休憩空间内布置商业、餐饮和娱乐休闲功能。

基于以上考虑，我们提出通过与基地周边四条轨道交通车站相通的中央地下广场和连接各个街区的"绿心"，将明确分开的区域自然而有机地结合起来的规划方案。形成完整统一的同时，旨在打造能够提供丰富的城市生活的城市空间。

考虑到功能上的连续性，在陆家嘴地区所在的东侧布置办公楼，在梅园社区所在的西侧布置住宅。将配套设施设在便于进出建筑裙房的位置。住宅街区与平台将一部分配套功能有效地连接起来。同时，将一部分酒店布置在景观视线优越的中央标志性建筑的高层。

总建筑面积	850,000 ㎡
设计期间	2008/7 (竞赛)

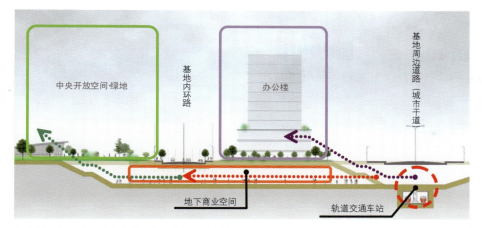

立体地处理从地铁站出发的动线，使其简洁明快

Shanghai Pudong Lujiazui area is the representation of Shanghai rapid development and the world-renowned tourist attractions. Along with the development of financial street, we seek to build a three-dimensional solution with humanistic block, work and life of equal importance which is different from the previous large blocks.

For this, we develop the following three main design goals:
1. Core area with high added value: a core area with high added value will be formed where lots of multinational companies in financial institutions, real estate, investment, consultation etc. are located.

2. Business district with convenient traffic: to ensure easy access for vehicles to the plot and parking. The public transport such as rail transport, buses and pedestrians routes to each plot are improved. Underground space is fully used; a pedestrian network will be formed in a short time which can disperse the flow of population.

3. High-quality comprehensive functional areas: improve the functions of exhibition, culture, hotels etc.; provide high quality supporting facilities for financial centers. The living functions are enriched; enhance the prosperity and charm of the comprehensive functional areas; the open space in the building podium and the plot are connected as a network; the commercial, catering and recreation functions are laid out in the colorful, prosperous and comfortable leisure space.

Based on the above considerations, we propose that by the central underground plaza which connects the rail stations of the four rail transports around the base and the "green heart" connecting the various blocks, we will clarify the planning program in which the separated areas are combined naturally and organically. While a complete uniform is formed, we aim to create a city space which can provide rich urban life.

Taking the continuity of functions into account; office buildings are arranged in the east of Lujiazui area; residential areas are arranged at the west of the plum community. The supporting facilitate are designed for easy access to the building podium. Residential blocks and platform will effectively connect with the supporting functions. Meanwhile, some hotels will be arranged in the high-rise of the central landmark building with superior landscape sight.

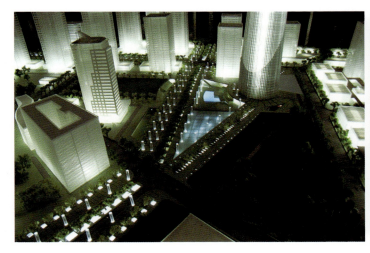
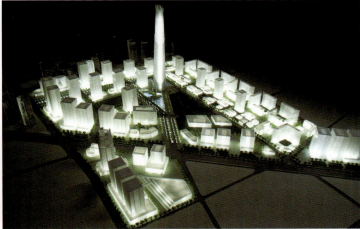

苏州工业园区城市设计
Suzhou Industrial Area Urban Planning

总建筑面积	850,000㎡
设计期间	2009
获奖经历	国际设计竞赛第一名

为了打造全新的CBD，我们提出营造富有特色的城市景观与实现高效的空间利用的设计方案。

我们的设计目标是：作为高速成长的CBD，与周边自然融为一体，环境优美宜人，在城市中能够亲身感受到"流水·绿色·天空"。
为此，我们提出以下三个设计提案。
【提案1】：形成个性独特的城市空间
【提案2】：塑造立体型城市
【提案3】：保证与周边地块的协作

具体的实施内容如下：
- 地理位置与苏州老城区相对，高层建筑的布局强调连接老城区与新工业园的轴线，打造地标性景观。
- 将各栋建筑有效地连接起来的地下广场和平台呈S形布局，同时兼顾功能性与富有特色的形态。

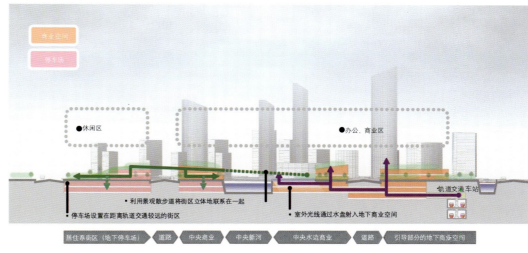

- 在中央布置利用基地的特色河流打造的亲水空间，一部分与上部铺有水盘的地下空间相连，形成易于辨识的开放的地下广场。
- 确保了明确的分区与流畅的动线，实现人车分离的同时，形成不同于周边连续开发的富有特色的城市空间。

扩大地区中央的自然水面，形成水系空间。使其成为基地整体中心，营造具有统一感的城市空间。另外，扩大水面空间：
- 可以增加面对水系的商业空间。
- 因为有仰望高楼的空间，特别是夜间照明的映射等，可以进行景观变化。
- 和地下空间的连接，考虑可以进行主题活动等各种各样的利用方法，为城市增加娱乐性、节日性等效果。

街区构成的剖面图
- 由于从轨道交通车站进入基地北侧的人流量很大，因此在地下设置商业空间，形成引导人流的魅力空间。
- 在街区中央设置充分利用水景的商业空间。
- 街区南侧作为休闲区，有一部分地形隆起，不仅形成与办公功能上的巧妙分割，还确保了一定程度的停车场。
- 酒店的休息厅位于200m的高空，将成为园区来访者的"空中休息厅"。
- 在可以俯瞰规划用地的中心和ＣＢＤ城市轴的西侧设有中庭大空间，可演绎出戏剧般的梦幻效果（例如结婚仪式等）。

To create a new CBD, we propose a design scheme which can create a distinctive urban landscape and space with efficient use.

Our design goal is a CBD with high-speed growth」 blended with the surrounding nature with beautiful and pleasant environment;「water flow・green・sky」can be experienced personally in the city.
For this, we propose the following three schemes:
【Scheme 1】:Urban space with unique characters will be formed.
【Scheme 2】:Create a three-dimensional city.
【Scheme 3】:To ensure the collaboration with surrounding land.

Implementation shall be as follows specifically.
· Its geographical location is opposite to the old Suzhou city. The layout of high-rise building stresses the axis which connects with the old city area and the new industrial park, creating a landmark landscape.
· The underground plaza and platform which link all the buildings effectively are S-shaped; meanwhile, take the functional and distinctive forms into account.
· A water space is created by using the distinctive river on the bases of the center. A part is connected with the underground space where a water pond is laid on the upper part to form a readily identifiable and open underground square.
· To ensure a clear division and smooth route, achieve the separation of people and vehicles; meanwhile, a distinctive urban space, which is different from the continuous development of its surrounding, is formed and highly appreciated.

Expand the natural water area in the center of the region; a prosperous river space is formed. Therefore, it will become the overall center of base to create an urban space with unified sense. In addition, expand the water space.
· The commercial space in water can be increased.
· Landscape change can be achieved because of the space from which we can look up to high buildings especially the representation in night lighting.
· For the connections with underground, a variety of methods can be used for theme activities to increase entertainment and festive effects etc.

The profile for block constitution
· Because there is huge flow of population from rail transport station to the north of the base, a commercial space is laid underground, as a charming space for leading the flow of population.
· A commercial space which fully uses the water landscape is laid in the center of the block.
· The south block is a recreation area with parts of uplifted land. Not only the ingenious partitions are achieved in the function of office areas, but also a parking lot can be ensured to a certain extent.
· The hotel lounge is located in the space with altitude of 200m. It will become a「sky lounge」which the park visitors can enjoy in the future.
· A large patio space is arranged in the center from which the planned land can be overlooked and the west of the city axis of CBD. A dramatic dream effect can be deduced. (For example, marriage etc.)

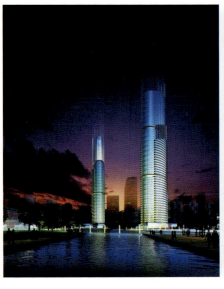

潍坊安顺新区整体规划及奥林匹克中心建筑概念设计
Olympic Center Conceptural Architecture Design and Anshun New Town Urban Planning, Weifang

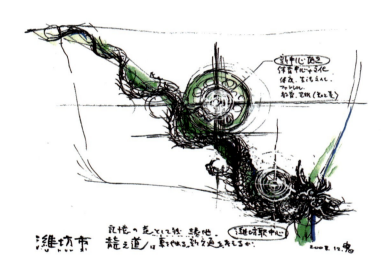

总建筑面积	154,997㎡
设计期间	2008
建设期间	2009~
合作设计	六角鬼丈设计工房
获奖经历	国际设计竞赛一等奖

潍坊安顺新区整体规划是日本设计在2008年取得了国际规划竞赛第一名后，进一步得到政府委托进行了奥林匹克中心建筑概念设计以及周边城市设计。本次规划方案的演绎取龙头风筝为设计母体，并在各个方面借鉴了其主要特点。

空间结构
在空间上，取其节节相扣的组织方式，形成城市公共中心有机串联的整体格局，延续了潍坊市整体空间形象。

组织功能
这条相互独立而又相互联系的空间轴线，涵盖了安顺新区的主要公共服务职能。包括体育中心、文化中心、商业中心、商务办公区、休闲娱乐区以及其他服务功能。

发展机制
龙首位于城市中心，龙身形成一系列"功能节"，其中体育中心位

于潍坊市风筝的龙尾,也是安顺区龙身的核心位置。规划借助文化体育中心形成的龙身率先发展,带动潍坊市的城市之鸢整体飞起。

功能定位

规划确定安顺新区的功能定位为:以"文化体育"为主题,辐射力量覆盖城市的西部郊区的城市副中心、区域级物流中心及城市西翼的重要门户。

城市品牌

面对全球化的挑战,国内市场经济条件下的城市规模扩张不可避免地将城市引入一场城市与城市之间的竞争。此时,城市品牌化的力量就是让人们了解潍坊,同时也让潍坊的纸鸢文化走入人们的视野。安顺新区体育文化中心是这个品牌的重要创造者之一。

奥体中心的概念建筑设施是核心区的重点项目。规划形成围绕由奥林匹克广场、观光塔和草坡绿化为核心的圈层结构,并通过三条轴线使内外有机联系。内层是核心功能层,包括体育比赛馆群和会展建筑群,延续了"龙鸢"的城市空间意象;外层是为内部服务的圈层,主要功能有综合训练馆、体校、停车楼、SOHO办公、大型商场和星级酒店等。

时代空间轴线联系观光塔、奥林匹克广场及集散广场中心,形成开敞、通透的视觉效果;体育场、网球中心和综合训练馆呈直线南北向布局,构成体育健康轴线,并在轴线上设置水景、绿化和雕塑造型,同时通过弧形的迥游体验轴线串联起体育场馆群建筑和会展建筑,并在两侧设置不同形式的景观及城市信息站,形成"水"和"绿"回游式大庭院。

The overall planning of Weifang Anshun New Area is designed by Japanese; after it won the first prize in the international planning competition in 2008, then it was entrusted by government to conduct concept design for the Olympic center construction and its surrounding urban design.

The interpretation of this scheme takes the dragon head kite as motif and learns its main features in all aspects.
Spatial structure
In spatial structure, interlocking organization is taken; an organic linked whole pattern of urban public centers is formed; it continues the whole spatial image of Weifang City.
Functional organization
This independent but interconnected spatial axis covers the main public service functions of Anshun New area, including sports center, cultural center, business center, business office area, leisure and entertainment area and other services.
Development Mechanism
Dragon head is located in the center of the city; the dragon body is with a series of "functional sections", among which Sports
The interpretation of this scheme takes the dragon head kite as motif and learns its main features in all aspects.
Spatial structure
In spatial structure, interlocking organization is taken; an organic linked whole pattern of urban public centers is formed; it continues the whole spatial image of Weifang City.
Functional organization

This independent but interconnected spatial axis covers the main public service functions of Anshun New area, including sports center, cultural center, business center, business office area, leisure and entertainment area and other services.
Development Mechanism
Dragon head is located in the center of the city; the dragon body is with a series of "functional sections", among which Sports Center is located in the dragon's tail of dragon head kite in Weifang City, and is also the core of the dragon body in Anshun new area. It is planned that the dragon body of cultural and sports center take the lead in development, then the eagle of Weifang City will fly wholly.
Functional position
It is planned that the functions of Anshun new areas are positioned as: themed as "cultural sports"; its radiation covers the vice city center in the western city, regional logistics center and an important gateway to the west wing of the city.
City Brand
Faced with global challenges, the urban scale expansion under the domestic market economy will inevitable lead cities into a competition between cities. At this point, the power of city brand is to make people understand Weifang; meanwhile the paper culture of Weifang also enters people's vision. The Sport and Culture Centre in Anshun New Area is one of the important creators of the brand.

The concept building facilities in the Olympic center are the key projects in the core area.
It is planned that the ring structure with Olympic Plaza, observation tower and grass

vegetation as core will be formed. The inside and outside are linked organically by the three axes. The inside is core function circle, including sports competition stadiums and convention groups, interconnected like dragon's eggs and continuing the urban spatial intention "Dragon and eagle"; the outer is a circle for internal service, with main functions such as comprehensive training hall, sports school, parking building, SOHO offices, and large commercial and star hotels.

The time spatial axes contact observation tower, Olympic Plaza and distribution square center, forming open, transparent visual landscape effect; stadium, tennis center and comprehensive training center are laid out in a line from north to south, formed as the axis of physical health; water, vegetation and sculpture are set on the axis; meanwhile, by circuitous travel, one can experience the stadium building groups and convention construction interconnected by axis; different forms of landscape square and city information stations are set on both sides; the large tour style courtyard with "water" and "vegetation" are formed.

长春高新开发区城市设计
Hi-Tech Center Urban Planning, Changchun

总建筑面积	2,776,300 ㎡
设计期间	2007/3~2008/1

长春高新开发区核心区位于长春南北新城的西部，是长春市城市南移的重要功能组成部分。核心区的规划结构"三轴一绿岛"刻画出主要城市结构关系：从核心区的A区开始形成商务轴、创意轴以及景观轴，在南侧形成人工绿岛，塑造高附加值的山体生态别墅等。

两轴包括一条沿着从长春市中心区延伸过来的主干道硅谷大街的"商务轴"，一条沿着绿色洋溢的景观道路超越大街的"创意轴"，在两轴之间，设置连接ABC地块的隐藏的"景观轴"。在A区的商务功能、B区的扩展服务区和C区的生活区互相联动协作。

核心区的区域中心是全新"绿岛"——缓缓隆起的绿岛台地，它是区域的中央公园，是文化产业中心

区的象征，其中分布着博物馆、体育设施、训练设施、会所，将所有功能有机地凝聚在一起，对市民开放。

在穿过核心区中心的景观轴上，设置地区的咽喉——管理委员会大厦、高层办公大楼、绿地广场。人流集中的建筑和可以疏散人流的广场相邻，形成环境优美的都市化的地区骨架。
在左翼的商务轴上，设置以连接地铁的地下商业街为基础的大型商业设施和办公建筑群。发挥临近车站的特点，形成充满活力的街区。

在右翼的创意轴上，设置酒店、SOHO等。提高商住混合的区域性，形成昼夜通明的安全街区。南侧将以绿岛为中心形成与环境一体化的居住区域设计。

Changchun High-tech Development Area is located in the core area of the West Changchun High-tech Development Area, Changchun North-South New City; it is an important function part of Changchun City Shift towards south. The planning structure "3 axes with a green Island" in the core area depicts the main urban structure relationship: business and creative axis and landscape-axis are formed from core Area A; an artificial green island is formed in the south to shape mountain ecological villas with high added value.

The two axes are: One is the [business-axis] of main Silicon Valley road extended from the downtown area of Changchun; the other is the [creative axis] beyond street along the landscape road full of green. A hidden [landscape axis] is set to connect ABC block. The business functions in area A, expansion service areas in Area B and living areas in Area C are collaborated with each other.

The regional centre in core area is grand new "green island"; it is the central park in the region symbolizing cultural industry center, where museums, sports facilities, training facilities, clubhouse are dotted. All the functions are gathered together organically, and open to the public.

On the landscape axis across core area, there are the throat – the Management Committee Building, senior office buildings and vegetation square of the region. The buildings with concentrated population are adjacent to the plaza where stream of people can be evacuated; an urbanization regional framework with beautiful environment is formed.

威海金线顶城市规划
Jinxianding Urban Planning, Weihai

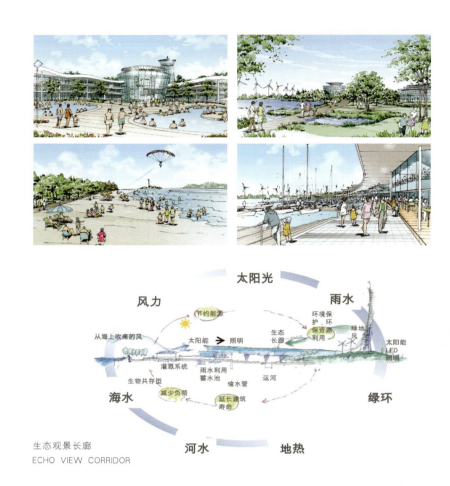

生态观景长廊
ECHO VIEW CORRIDOR

总建筑面积	1,230,000 ㎡
设计期间	2008/06~2008/10
建设期间	2008/12~2011/11
竣工年月	2011/12
获奖经历	国际设计竞赛第一名

威海市金线顶地段整体改造项目是国际竞赛第一名项目,目前进入了紧张的实施工程。

金线顶地区以"填海造地,自然再生"为目标,将起到改善和提高威海自然生态环境质量的作用。新区地面由北到南景色各异,建筑高度逐渐升高:从42m的金线顶山,再到南侧沿海展开的多层观海住宅,到高达150m的高层酒店式公寓,形成了低、中、高布局,同时与开发容积率吻合。

金线顶区域的整体规模是100hm²,其中将通过填海造地开发47hm²的土地,包括新建的16hm²绿地和9hm²的海滩。

整个用地分为5个区域:
功能区——海警和国际客运码头

海洋观光休闲区——海滨区域是各种设施和自然景观、填海的新生景观巧妙融

合的区域。区域内设置了客运码头和市内乃至国内最高水准的水族馆,将成为特色的标志性区域。

金线顶山地区——绿化区结合金线顶原有山体,改造原有山上的部分建筑,恢复自然景观。一条景观生态长廊巧妙地结合地形,将绿色山体和生态水生植被区结合。

商业文化区——城市轴线依托高架道路,面对威海之海韵广场,将城市发展与大海连接一起,极具动感和魅力

生态居住区——生态居住区位于区域南侧,包括超高层、高层、多层、别墅等不同的形态和居住组团。所有住宅都突出了和水空间和绿化空间的结合。

自然轴线——生态观景长廊从金线顶指向大海,刘公岛,成为视线景观的延续,体现了自然地域的象征性。开放空间、广场的亲和力和建筑彼此相互依托的魅力,形成了具有独特情结的新威海娱乐休闲商业中心,成为具有标志性的核心区开发模式

金线顶规划紧紧围绕日本设计的全新设计理念——环境城市的概念，积极推动绿色生态规划：
1. 创造可以与大海·大气·土地共生的可持续性发展的城市
2. 导入污水处理设施
3. 可再生能源利用
4. 杜绝污染大海
5. 建设区域能源中心，有效利用南北区域建筑使用特点，通过区域能源中心改善区域内部能源分布和利用，统筹调配，减少资源浪费
6. 能源循环系统
7. 废弃物回收系统
8. 防风措施
9. 控制刮向地表面的风
10. 防波措施
11. 体现海洋特色的规划：生态公园和人造运河

Weihai City, the overall transformation project of Jinxianding Urban Planning is the first prize winner project; after coordinating with various related units, now we start the tense implementation of the project.

Jinxianding area is targeted as "land reclamation, natural regeneration", which will serve to improve and enhance the quality of natural environment in Weihai. The ground scenery in the new area varies from north to south, its building height gradually increases: from 42-m Jinxianding hill to the multi-storey sea-view residential apartment in the south coastal, 150 m high-rise hotel apartments, form a low, middle and high layout, meanwhile it is consistent with the development floor area ratio.

The overall size of Jinxianding area is 100 hectares, among which 47ha land will be developed through reclamation of land, including the new 16ha vegetation and 9ha beaches.

The whole land is divided into five regions:
Functional areas — marine police and the International Ferry Terminal.

Marine tourism and leisure area — the coastal area is an area where a variety of facilities and natural landscape, the new reclamation area landscape are integrated in a smart way. Passenger terminal and aquarium which are of the highest level in the city or China are set in the region; so this area will become the landmark with characteristics.

Jinxianding area — the vegetation area combines with the original mountains of Jinxianding; transform parts of the buildings on the original hills, restore the natural landscape. A landscape ecological corridor combines with terrain in a clever way; the green mountain and the ecological aquatic vegetation are combined.

Commercial and cultural area — the city axis relies on elevated road. Facing the Haiyun Square of Weihai, we will combine urban development with the sea; they are very dynamic and charming.

Ecological residential area — the ecological residential area is located in the south of region, including super high-rise, high-rise multi-floor, villas and other different forms and residential groups. All the households highlight the combination of vegetation space and water space.

Natural axis — From Jinxianding, the ecological landscape corridor points to the sea and Liugong Island, as a continuation of the landscape view. This reflects the symbolism of natural area. The charm of open space, plazas affinity and construction relying on each other form an entertainment commercial center with unique complex in the new Weihai. It becomes a symbolic development model in the core area.

Jinxianding planning is around the new design concept of Japanese design – the concept of environment city, and actively promotes the vegetation ecological planning:
1. Create a city with sustainable development which is in symbiosis with sea, air and land
2. Introduce sewage treatment facilities
3. Use renewable energy
4. To prevent sea pollution
5. Build a regional energy center: apply the characteristics of north-south regional buildings efficiently; by the regional energy center, we can improve the distribution and use of energy within the region, co-ordinate and reduce the waste of resources.
6. Energy circulation system
7. Waste Recycling System
8. Windbreak measures
9. Control the wind which blows towards the ground surface
10. Anti-wave measures
11. Planning reflecting marine features: ecological park and man-made canal

图书在版编目（CIP）数据

日本设计 NIHON SEKKEI：汉英对照/株式会社
日本设计编著. --北京：中国建筑工业出版社，2011.12
ISBN 978-7-112-13788-6

Ⅰ．①日… Ⅱ．①日… Ⅲ．①建筑设计-日本-现代-图集
Ⅳ．①TU206

中国版本图书馆CIP数据核字（2011）第235831号

责任编辑：戴 静　丁 夏
装帧设计：深圳市品筑文化传播有限公司

日本设计 NIHON SEKKEI
日本设计编委会：叶晓健　生田惠里子　安藤一将　Henry Tsang　金在虎　唐君
*
中国建筑工业出版社出版、发行（北京西郊百万庄）
各地新华书店、建筑书店经销
利丰雅高（深圳）有限公司制版
利丰雅高（上海）有限公司印刷
*
开本：965×1270毫米 1/16　印张：14½　字数：200千字
2011年12月第一版　2011年12月第一次印刷
定价：180.00元
ISBN 978-7-112-13788-6
　　　（21568）

版权所有　翻印必究
如有印装质量问题，可寄本社退换
（邮政编码 100037）